Probability and Statistics

for Cambridge International A Level

J. Chambers • J. Crawshaw • P. Balaam

OXFORD

UNIVERSITY PRESS

S1 Contents

Introduction

The *Mathematics for Cambridge International A Level* series has been written specifically for students of Cambridge's 9709 syllabus by an experienced author team in collaboration with examiners who are very familiar with the syllabus and examinations. This means that, no matter which combination of modules you have chosen, the content of this series matches the content of the syllabus exactly and will give you firm guidelines on which to base your studies.

In this book, the content of the Statistics 1 module is divided into 6 chapters that give a sensible order for your studies. Each chapter begins with a list of objectives showing you exactly what is covered.

The following features help you to understand the concepts of the S1 module and to succeed in your exams.

- The introductions to concepts are accompanied by examples of questions, many of which are actual Cambridge questions, together with their solutions. These show each step of working along with a commentary on the reasoning processes involved.

- There are numerous exercises for you to practise what you have learned and develop your skills.

- At the end of each chapter there is a Mixed Exercise with more detailed questions covering the content of the chapter. These questions are similar to those found in exam papers and many are from real exam papers.

- Summaries of key information and formulae are given at the end of each chapter to help you revise what you have covered in the chapter.

- Answers to all questions are provided at the back of the book for you to check your answers to exercises.

- Two sample exam papers have been created in the style of Cambridge's International A Level Exams S1 to give you some experience of working a full exam paper.

1 Representation of data

In this chapter you will learn about

- types of data
  - qualitative
  - quantitative: discrete, continuous

- diagrams to illustrate data
  - bar charts, pie charts
  - vertical line graphs
  - stem-and-leaf diagrams
  - histograms
  - cumulative frequency graphs
  - box-and-whisker plots

- averages (measures of central tendency)
  - mode, median, mean

- spread (measures of variability)
  - range, interquartile range, standard deviation

- advantages and disadvantages of different measures

REPRESENTING DATA

In everyday life you receive much factual information from the internet, newspapers and television. The question is, how do you make sense of it?

The information, particularly when it comes from surveys and experiments, is a collection of observations, known as data. The branch of mathematics dealing with the collection, analysis and interpretation of numerical data is called Statistics.

When you collect data in an experiment or a survey it is difficult to interpret it in its raw form. This chapter looks at ways to present and summarise the data so that it is easier to analyse, such as

- ordering the data in a table or chart
- drawing a diagram to provide a quick visual snapshot or to estimate values
- working out an average to represent the data, known as a measure of central tendency
- working out a measure of the variability, or spread, of the data.

TYPES OF DATA

There are two types of data: qualitative and quantitative.

Qualitative data

Qualitative data consists of descriptions, using names, for example,
- colours of cars sold in a garage during a month
- varieties of apple trees in a large orchard
- types of vehicle using a particular road in an hour.

Qualitative data are often represented visually in a bar chart or a pie chart where it is easy to see the most common category.

Example 1.1

The table shows the areas of different categories of land use in a particular region.

Category of land use	Urban	Woodland	Farmland	Reservoirs	
Area (km²)	615	660	1200	225	Total 2700

Illustrate the data diagrammatically.

You could draw a pie chart.

Working for angles in pie chart:

Total area of the region $= 2700\,\text{km}^2$

Urban: $\dfrac{615}{2700} \times 360° = 82°$

Woodland: $\dfrac{660}{2700} \times 360° = 88°$

Farmland: $\dfrac{1200}{2700} \times 360° = 160°$

Reservoirs: $\dfrac{225}{2700} \times 360° = 30°$

Pie chart to show land use

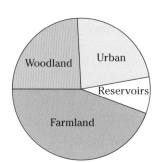

Alternatively you could draw a bar chart.

Note that in a bar chart
– all the bars have the same width
– there are distinct gaps between the bars for each category.

Bar chart to show land use

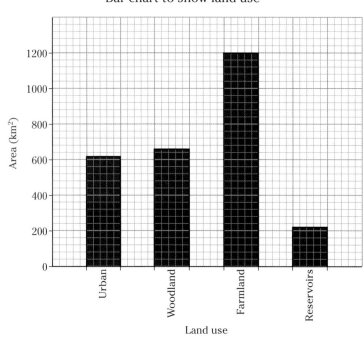

Bar charts can also be drawn to **compare** sets of data.

The table below shows, in millions of dollars, the sales of a company in two successive years.

Sales ($ million)	Africa	America	Asia	Europe	
Year 1	5.5	6.7	13.2	19.6	Total sales $45 million
Year 2	5.8	15.2	9.2	29.8	Total sales $60 million

To illustrate the data you could draw a **comparative bar chart**:

The bars can be horizontal or vertical in a bar chart.

Comparative bar chart to show sales in Year 1 and Year 2

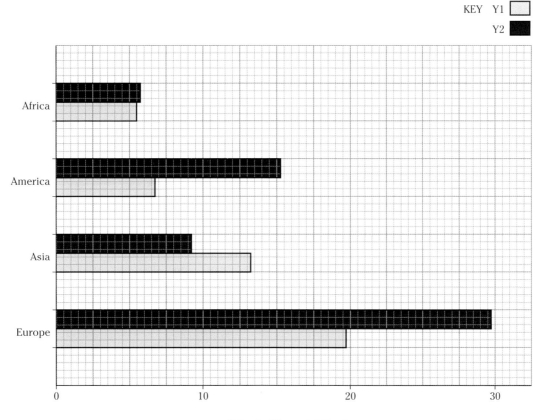

Sales (millions of dollars)

Exercise 1a

1 A school recorded the number of papers awarded each grade in the A Level examinations. The results were:

A*: 44, A: 88, B: 69, C: 30, D: 17, E: 4.

Represent the data on a bar chart.

2 The following data summarise the expenditure by a local council during a particular year.

Service	Expenditure ($ million)
Education	160.2
Highways	35.7
Police	28.9
Social Services	27.9
Other	24.5

These data are to be represented by a pie chart of radius 5 cm. Calculate, to the nearest degree, the angle corresponding to each of the five classifications.
(There is no need to draw the pie chart)

3 There are 24 pupils in Peter's class. He carried out a survey of how the pupils in his class travelled to school. His results are shown in the table.

Method of travel	Number of pupils
Bus	10
Car	2
Bicycle	5
Walking	7

The data are to be illustrated by a pie chart.

(i) Calculate, to the nearest degree, the sector angles of the pie chart.

(ii) Draw the pie chart using a circle of radius 4 cm, labelling each sector with the method of travel that it represents.

4 There are three categories of tickets for concerts at a particular venue: Child, Adult and Senior. A survey of the tickets sold for two consecutive performances produced the results shown in the table.

Percentage of tickets sold in each category

	Child	Adult	Senior
Performance 1	20	30	50
Performance 2	35	55	10

Illustrate these results diagrammatically.

5 A golf club has 180 members. There are three categories of membership: men, women and juniors. There are three times as many men as women and half as many juniors as women.

Illustrate the information using

(i) a bar chart,

(ii) a pie chart.

6 A charity obtains its income from various sources. The table shows these sources and the corresponding amounts of income for last year.

Source	Income ($)
Advertising	30 000
Donations	x
Fees	9000
Investment	3000
Sponsorship	10 000

A pie chart was drawn to illustrate the data. Given that the angle of the sector representing Donations was 204°, calculate the value of x.

Quantitative data

Quantitative data take numerical values. There are two types of quantitative date: discrete and continuous.

Discrete data

In a survey on the number of letters in the solutions of a crossword puzzle, the following data were obtained from the crossword puzzle in Monday's newspaper.

9	5	7	7	5	12	6	6	12	5
7	7	5	5	3	7	3	7	9	6
3	5	7	3	6	7	7	6	3	7

This is an example of **discrete** data.

Discrete data can take only **exact** values, for example,

- the number of rally cars passing a checkpoint in a ten-minute interval
- the number of hits on an internet website in an hour
- the shoe sizes of children in a class
- the number of goals scored in a football match.

Discrete data often arise from counting.

Frequency distribution for discrete data

To illustrate discrete data concisely, count the number of times each value occurs and summarise these in a table, known as a **frequency distribution**.

This is the frequency distribution for the letters in the solutions of the crossword puzzle.

Number of letters in word	3	5	6	7	9	12	
Frequency	5	6	5	10	2	2	Total 30

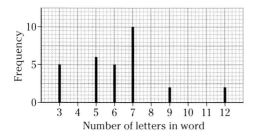

The data can be illustrated in a vertical line graph.

In a **vertical line graph** the height of the line is equal to the frequency.

The mode

The **mode** is the value that occurs most often.

The mode is the most popular value, deriving from the French 'à la mode' meaning fashionable.

The mode is a useful **average** (typical value) when there is a single most common value or perhaps two (bi-modal). However, it is not very helpful in summarising the data if there are more than two modes.

In the example above:

The most common length of word in the crossword puzzle is 7, so the mode is 7 letters.

Make sure that you give the most common value of the variable, not the frequency with which it occurred.

Continuous data

The following data were obtained in a survey of the heights of 20 children in a sports club. Each height was measured to the nearest centimetre.

> 133   136   120   138   133   131   127   141   127   143
>
> 130   131   125   144   128   134   135   137   133   129

This is an example of **continuous** raw data. The heights are given to a specified degree of accuracy. For example, a height recorded as 144 cm, to the nearest cm, could have arisen from any value in the interval 143.5 cm ≤ height < 144.5 cm.

Continuous data cannot take exact values but can be given only within a specified range or to a specified degree of accuracy.

Other examples of continuous variables are

- the times taken by 10-year-olds to complete a puzzle
- the masses of potatoes in a bag
- the speeds of vehicles passing a checkpoint.

Continuous data usually arise from measuring quantities.

Frequency distribution for continuous data

To form a frequency distribution of the heights of the 20 children, group the information into **classes** or **intervals**. Here are different ways of writing the same set of intervals.

Height (to the nearest cm)	Height (to the nearest cm)	Height (cm)	Height (cm)
120–124	$120 \leqslant h \leqslant 124$	$119.5 \leqslant h < 124.5$	119.5–124.5
125–129	$125 \leqslant h \leqslant 129$	$124.5 \leqslant h < 129.5$	124.5–129.5
130–134	$130 \leqslant h \leqslant 134$	$129.5 \leqslant h < 134.5$	129.5–134.5
135–139	$135 \leqslant h \leqslant 139$	$134.5 \leqslant h < 139.5$	134.5–139.5
140–144	$140 \leqslant h \leqslant 144$	$139.5 \leqslant h < 144.5$	139.5–144.5

The values 119.5, 124.5, 129.5, ... are called the **class** (or **interval**) **boundaries**. The upper class boundary of one interval is the lower class boundary of the next interval.

Width of an interval

The **width** of an interval is the difference between the boundaries of the interval.

> Interval width = upper boundary − lower boundary.

So, the width of the first interval (120–124) is $124.5 - 119.5 = 5$.

Take care: the width is **not** 4, as the boundaries are not 120 and 124.

In the tables, the intervals have been chosen so that they all have width 5.

To group the heights it helps to use a tally column, entering the numbers in the first row 133, 136, 120, ... etc and then the second row. It is a good idea to cross off each number in the list as you enter it.

This is the frequency distribution for the above data:

Height (cm)	Tally	Frequency
$119.5 \leqslant h < 124.5$	│	1
$124.5 \leqslant h < 129.5$	ЖＩ	5
$129.5 \leqslant h < 134.5$	ЖＩ ‖	7
$134.5 \leqslant h < 139.5$	‖‖‖	4
$139.5 \leqslant h < 144.5$	‖‖	3
		Total 20

It is important to note that when the data are presented only in the form of a grouped frequency distribution, the original information has been lost. For example, you would not know the value of the item in the first interval, only that it is somewhere in the interval $119.5 \leqslant h < 124.5$.

STEM-AND-LEAF DIAGRAMS

A way of grouping data into intervals while still retaining the original data is to draw a **stem-and-leaf diagram**, also known as a **stemplot**.

These are the marks of 20 students in an assignment:

84   17   38   45   47   53   76   54   75   32

66   65   55   54   51   44   39   19   54   72

Notice that the lowest mark is 17 and the highest mark is 84.

In stem-and-leaf diagrams all the intervals must be of **equal width**, so it seems sensible to choose intervals 10–19, 20–29, 30–39, …, 80–89 for this data, so take the **stem** to represent the tens and the **leaf** to represent the units.

This is how you would start: The first five entries 84 17 38 45 47 are represented like this. Cross the numbers off the list as you go.

Stem (tens)	Leaf (units)
1	7
2	
3	8
4	5 7
5	
6	
7	
8	4

When all the numbers have been entered, the diagram should look like this.

Preliminary plot

Stem	Leaf	
1	7 9	(2)
2		(0)
3	8 2 9	(3)
4	5 7 4	(3)
5	3 4 5 4 1 4	(6)
6	6 5	(2)
7	6 5 2	(3)
8	4	(1)
		Total 20

It is useful to count the number of entries in each row. These are not part of the stem-and-leaf diagram, however, and will not be required in the examination.

Check that the total is 20.

Now arrange the entries in each row in numerical order with the smallest next to the stem.

Important: You must always give a **key** to explain what the stem and leaf represent.

Final plot

Assignment marks

1	7 9	
2		
3	2 8 9	
4	4 5 7	
5	1 3 4 4 4 5	
6	5 6	
7	2 5 6	
8	4	

Key: 1 | 7 means 17 marks

The stem-and-leaf diagram gives a good idea, at a glance, of the shape of the distribution. It is easy to pick out the smallest and largest values (17 and 84) and to see that the mode is 54.

Note that the row with stem 2 must be left in the diagram, even though it has no entries.

In a stem-and-leaf diagram
  – equal intervals must be chosen
  – a **key** is essential.

Example 1.2

The lengths, in metres, of 20 measurements in a physics experiment are recorded as follows.

1.78, 1.87, 1.89, 1.72, 1.68, 2.04, 1.96, 1.76, 1.90, 1.73,
1.78, 1.61, 1.78, 1.77, 1.85, 1.65, 1.89, 1.95, 2.01, 1.83

(i) Represent this information on a stem-and-leaf diagram.

(ii) State the mode.

(i) The lowest value is 1.61 and the highest value is 2.01.

Remember to include the key.

Lengths

16	1 5 8
17	2 3 6 7 8 8 8
18	3 5 7 9 9
19	0 5 6
20	1 4

Key: 17 | 3 means 1.73 m

(ii) The mode is 1.78 m.

Example 1.3

The maximum temperature in °C, measured to the nearest degree, was recorded each day during June in a particular city. The temperatures were as follows:

19 23 19 19 20 12 19 22 22 16 18 16 19 20 17
13 14 12 15 17 16 17 19 22 22 20 19 19 20 20

Draw a stem-and-leaf diagram to illustrate the temperatures and write down the mode.

(i) The lowest value is 12 and the highest value is 23.

If you group the data into the intervals 10–19, 20–29 this would give you very little information.

Choose a sensible number of intervals; usually between 5 and 10. Since you must use intervals with equal width, you could use intervals of 12–13, 14–15, 16–17, 18–19, 20–21, 22–23.

First do a preliminary plot and then arrange the entries in each leaf in order.

Maximum daily temperatures

1	2 2 3
1	4 5
1	6 6 6 7 7 7
1	8 9 9 9 9 9 9 9 9
2	0 0 0 0 0
2	2 2 2 2 3

Key: 1 | 2 means 12 °C

(ii) The mode is 19 °C.

Back-to-back stem-and-leaf diagrams

Stem-and-leaf diagrams can be used to compare two sets of data by showing them together on a **back-to-back stem-and-leaf diagram**.

Example 1.4

These are the examination marks for French and for English achieved by pupils in a particular class.

French	43 55 29 49 36 55 61 34 42 42 54 60 48 23 44 31 55 45 37 57
English	80 65 74 59 79 92 52 71 43 86 60 74 57 41 79 74 58 52 64 84

(i) Draw a back-to-back stem-and-leaf diagram.

(ii) Compare the two sets of marks.

(i)   Examination marks

French		English
	1	
9 3	2	
7 6 4 1	3	
9 8 5 4 3 2 2	4	1 3
7 5 5 5 4	5	2 2 7 8 9
1 0	6	0 4 5
	7	1 4 4 4 9 9
	8	0 4 6
	9	2

Key: 2 | 4 | 1 means 42 for French
41 for English

(ii) The lowest marks are 23 for French and 41 for English. The highest marks are 61 for French and 92 for English.

The marks for English are more spread out (more variable) than the marks for French.

The mode for French is 55 and the mode for English is 74.

It would appear that the pupils performed better in English. However, this would depend on the standards of marking used in the two examinations.

Exercise 1b

1 The children in a class were asked how many crayons they had in their pencil case. The vertical line diagram shows the results.

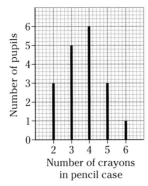

(i) How many children were there in the class?

(ii) If all the crayons were put into a big pot, how many crayons would there be?

2 The weights, correct to the nearest kg, of 30 men are shown below.

```
74  52  67  68  71  76  86  81  73  68
64  75  71  61  63  57  67  57  59  72
79  64  70  74  77  79  65  68  76  83
```

(i) Draw a stem-and-leaf diagram to show the data.

(ii) Write down the mode.

3 In a lesson on measurement, 30 pupils estimated the length of a line, writing their guess in cm, to the nearest mm, with the following results.

```
9.2  7.3  7.0  6.5  5.4  5.3  10.1  8.4  8.8
7.1  7.5  7.9  6.7  9.6  5.5   7.4  7.0  8.2
5.5  7.8  8.2  7.6  6.1  6.1   3.9  6.8  7.6
8.1  8.0  10.0
```

(i) Draw a stem-and-leaf diagram to show the data.

(ii) The line was 7.5 cm long. What percentage of pupils overestimated the length?

4 The daily hours of sunshine in a particular city during August were as shown.

```
7.0  7.6  12.5  12.9   8.3   9.7  8.4  11.1
7.5  7.5   9.8  10.4  11.6  11.3  7.3   7.8
6.8  6.2   6.1   5.6   5.6   5.8  4.8   4.3
0.0  0.6   0.8   1.6   0.2   2.4  2.6
```

Illustrate these data on a stem-and-leaf diagram.

5 Cartons are advertised as containing 2 litres of milk. In a quality control test, the volume of milk in 20 randomly selected cartons was measured in litres, to the nearest millilitre, and gave the following results.

2.017, 2.015, 2.019, 1.982, 2.003, 1.986, 2.024, 2.017, 2.001, 2.004, 1.988, 2.033, 1.990, 2.018, 2.011, 2.023, 2.019, 2.022, 2.008, 1.985

(i) Draw a stem-and-leaf diagram to show the data.

(ii) What percentage of cartons contained less than the advertised volume of milk?

6 These are the pulse rates of 30 company directors measured before and after taking exercise.

Before exercise: 110, 93, 81, 75, 73, 73, 48, 53, 69, 69, 66, 111, 105, 93, 90, 50, 57, 64, 90, 111, 91, 70, 70, 51, 79, 93, 105, 51, 66, 93.

After exercise: 117, 81, 77, 108, 130, 69, 77, 84, 84, 86, 95, 125, 96, 104, 104, 137, 143, 70, 80, 131, 145, 106, 130, 109, 137, 75, 104, 75, 97, 80.

(i) Draw a back-to-back stem-and-leaf diagram to show the data.

(ii) Compare the pulse rates before and after exercise.

7 These are the ages of teachers in two schools.

School A: 51, 45, 33, 37, 37, 27, 28, 54, 54, 61, 34, 31, 39, 23, 53, 59, 40, 46, 48, 48, 39, 33, 25, 31, 48, 40, 53, 51, 46, 45, 45, 48, 39, 29, 23, 37.

School B: 59, 56, 40, 43, 46, 38, 29, 52, 54, 34, 23, 41, 42, 52, 50, 58, 60, 45, 45, 56, 59, 49, 44, 36, 38, 25, 56, 36, 42, 47, 50, 54, 59, 47, 58, 57.

Draw a back-to-back stem-and-leaf diagram and compare the ages of the teachers in the two schools.

HISTOGRAMS

Grouped data can be displayed in a **histogram**, as in the following diagram:

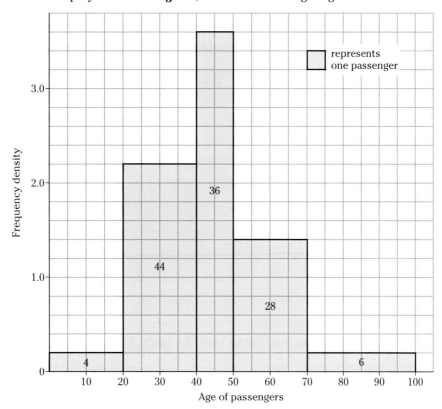

This histogram represents the following distribution of ages of passengers on a shuttle flight from Denver, Colorado to Salt Lake City, Utah.

Age, x years	$0 \leqslant x < 20$	$20 \leqslant x < 40$	$40 \leqslant x < 50$	$50 \leqslant x < 70$	$70 \leqslant x < 100$
Frequency	4	44	36	28	6

Histograms resemble bar charts, but there are two important differences:

There are **no gaps** between the bars.
The **area** of the bar is proportional to the frequency that it represents.

Histograms often have bars of different widths, so the height of the bar must be adjusted in accordance with the width of the bar.

If the height of the bar is adjusted so that the area is **equal** to the frequency in that interval

then   area = height × interval width = frequency,

so   height = $\dfrac{\text{frequency}}{\text{interval width}}$

This height is known as the **frequency density**.

So,   frequency density $= \dfrac{\text{frequency}}{\text{interval width}}$

Using frequency density:

  area of the bar = frequency in that interval

  total area = total frequency.

In the histogram above, the interval $20 \leqslant x < 40$ has frequency 44. The lower class boundary is 20 and the upper class boundary is 40, so interval width $= 40 - 20 = 20$.

  Frequency density $= \dfrac{\text{frequency}}{\text{interval width}} = \dfrac{44}{20} = 2.2$

This is the frequency table showing the calculation of the frequency densities.

Age	Class boundaries l.c.b	u.c.b	Interval width	Frequency	Frequency density
$0 \leqslant x < 20$	0	20	20	4	$\frac{4}{20} = 0.2$
$20 \leqslant x < 40$	20	40	20	44	$\frac{44}{20} = 2.2$
$40 \leqslant x < 50$	40	50	10	36	$\frac{36}{10} = 3.6$
$50 \leqslant x < 70$	50	70	20	28	$\frac{28}{20} = 1.4$
$70 \leqslant x < 100$	70	100	30	6	$\frac{6}{30} = 0.2$

Modal class

When data have been grouped it is not possible to state the mode. Instead the modal class can be given.

The **modal class** is the interval with the greatest frequency density. In a histogram it is the interval represented by the highest bar.

In the histogram above, the interval with the highest bar is $40 \leqslant x < 50$, so this is the modal class, indicating that people in their forties comprised the most common group.

Notice from the table that this interval has the greatest frequency density (3.6), but it does not have the greatest frequency.

Example 1.5

A survey on the duration of telephone calls made to an office on a particular day gave the following results:

Duration, t minutes	$1 \leqslant t < 3$	$3 \leqslant t < 9$	$9 \leqslant t < 15$	$15 \leqslant t < 20$
Frequency	10	42	12	7

Draw a histogram to represent the data.

For each interval, write out the class boundaries, work out the interval width and then calculate the frequency density.

Duration (minutes)	Class boundaries l.c.b	u.c.b	Interval width	Frequency	Frequency density
$1 \leqslant t < 3$	1	3	2	10	$\frac{10}{2} = 5$
$3 \leqslant t < 9$	3	9	6	42	$\frac{42}{6} = 7$
$9 \leqslant t < 15$	9	15	6	12	$\frac{12}{6} = 2$
$15 \leqslant t < 20$	15	20	5	7	$\frac{7}{5} = 1.4$

Histogram to show duration of telephone calls

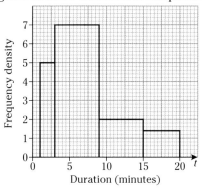

Example 1.6

The grouped frequency table records the weights, to the nearest gram, of the letters delivered to an apartment block on a particular day.

Weight (grams)	31–50	51–60	61–70	71–100	101–150
Frequency	16	25	36	33	10

Draw a histogram to illustrate the data and state the modal class.

For each interval, write out the class boundaries, work out the interval width and then calculate the frequency density.

The weights have been recorded **to the nearest gram**, so the interval 31–50 represents $30.5 \leqslant$ weight < 50.5. The lower boundary is 30.5 and the upper boundary is 50.5.

This is an example of grouped continuous data where the groups are described using rounded values (to the nearest gram). It could seem that there are gaps between the groups, but when you consider the class boundaries you will see that there are **no gaps**.

Weight (nearest gram)	Class boundaries l.c.b	u.c.b	Interval width	Frequency	Frequency density
31–50	30.5	50.5	20	16	$\frac{16}{20} = 0.8$
51–60	50.5	60.5	10	25	$\frac{25}{10} = 2.5$
61–70	60.5	70.5	10	36	$\frac{36}{10} = 3.6$
71–100	70.5	100.5	30	33	$\frac{33}{30} = 1.1$
101–150	100.5	150.5	50	10	$\frac{10}{50} = 0.2$

Histogram to show weights of letters

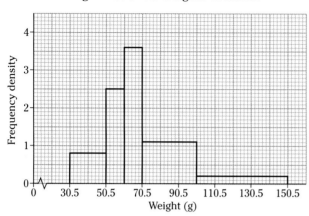

Remember to include a title.

Remember to label the axes.
Note that
– the vertical axis (frequency density) **must** start from zero.
– the horizontal axis need not start from zero.

HINT: You will find it easier to draw the histogram if you mark the class boundaries (30.5, 50.5, etc.) at the thicker lines on your graph paper.

In this example, the interval with the greatest frequency density (represented by the highest bar) is also the interval with the greatest frequency.

The modal class is 61–70.

Note: As the weights are measured to the nearest gram, the intervals could be written $31 \leqslant w \leqslant 50$, $51 \leqslant w \leqslant 60$, $61 \leqslant w \leqslant 70$, $71 \leqslant w \leqslant 100$, $101 \leqslant w \leqslant 150$. The class boundaries are still 30.5, 50.5, 60.5 etc.

Example 1.7

One evening a waiter measured the amounts of water left by diners in the bottles on the tables in a restaurant. The volumes were measured to the nearest millilitre.

Volume (nearest ml)	0–19	20–39	40–89	90–189
Frequency	10	8	12	20

(i) State the class boundaries of each interval and calculate the frequency densities needed to draw a histogram.

(ii) State the modal class.

(i) The interval 0–19 represents $0 \leqslant$ volume < 19.5 ml, so the interval width is 19.5, whereas 20–39 represents 19.5 ml $\leqslant$ volume < 39.5 ml, so the interval width is 20.

This is an example of continuous (rounded) data where the first interval starts at 0.

The boundaries, widths and frequency densities are shown in the table below.

Volume (nearest ml)	Class boundaries l.c.b	u.c.b	Interval width	Frequency	Frequency density
0–19	0	19.5	19.5	10	$\frac{10}{19.5} = 0.51$ (2 s.f.)
20–39	19.5	39.5	20	8	$\frac{8}{20} = 0.4$
40–89	39.5	89.5	50	12	$\frac{12}{50} = 0.24$
90–189	89.5	189.5	100	20	$\frac{20}{100} = 0.2$

(ii) The modal class is 0–19.

Special care must be taken with class boundaries when drawing a histogram to represent **grouped discrete data**. As there are no gaps between the bars in a histogram, a *continuous scale* is used to represent *discrete* quantities.

For example

- the integer 2 is represented by an interval from 1.5 to 2.5
- data grouped as 10–19 is represented by the interval 9.5 to 19.5.

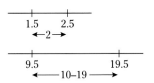

The difficulty comes, however, when the grouping starts at 0.

Consider the group 0–9. Although it may seem strange, when data are *discrete*, the lower boundary is taken as -0.5 and the group 0–9 is represented on a histogram by the interval from -0.5 to 9.5, as in Example 1.8 below.

Notice that the lowest class boundary is different from the interval described in Example 1.7 where the data are continuous.

Example 1.8

These are the marks in a statistics test for a group of 120 A level students.

Mark	0–9	10–19	20–29	30–49	50–79
Frequency	8	21	53	28	10

Represent the data on a histogram.

The marks are discrete. Take cake care with the boundaries.

Mark	Class boundaries l.c.b	u.c.b	Interval width	Frequency	Frequency density
0–9	-0.5	9.5	10	8	$\frac{8}{10} = 0.8$
10–19	9.5	19.5	10	21	$\frac{21}{10} = 2.1$
20–29	19.5	29.5	10	53	$\frac{53}{10} = 5.3$
30–49	29.5	49.5	20	28	$\frac{28}{20} = 1.4$
50–79	49.5	79.5	30	10	$\frac{10}{30} = 0.3$ (1 d.p.)

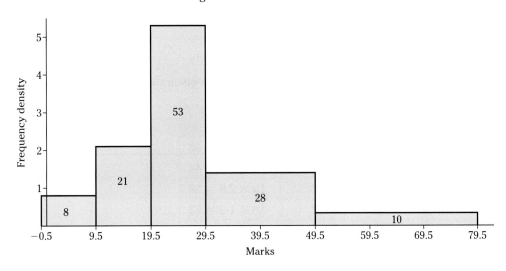

Histogram to show test marks

HINT: The vertical axis does not have to go through $(0, 0)$ but can be placed in a convenient position, for example:

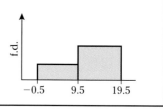

Finding frequencies from a histogram

The frequency in a particular interval is given by the area of the bar, so it is found by multiplying the interval width by the frequency density.

Frequency = interval width × frequency density

Example 1.9

A Passengers' Association conducted a survey on the lateness of trains arriving at a particular railway station. The results are illustrated in the histogram.

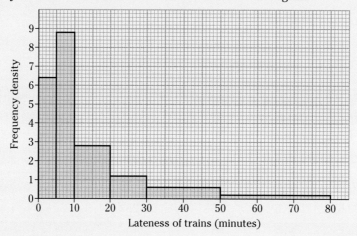

(i)   Construct a frequency table.

(ii) What percentage of the trains were less than 20 minutes late?

(i)   To find the frequency in each interval, use

You are finding the area of each bar.

frequency = interval width × frequency density

Calculate the interval widths and read the values of the frequency density from the histogram.

Lateness, t minutes	Interval width	Frequency density	Frequency
$0 \leqslant t < 5$	5	6.4	$5 \times 6.4 = 32$
$5 \leqslant t < 10$	5	8.8	$5 \times 8.8 = 44$
$10 \leqslant t < 20$	10	2.8	$10 \times 2.8 = 28$
$20 \leqslant t < 30$	10	1.2	$10 \times 1.2 = 12$
$30 \leqslant t < 50$	20	0.6	$20 \times 0.6 = 12$
$50 \leqslant t < 80$	30	0.2	$30 \times 0.2 = 6$

   (iii) Number of trains in survey

$$= 32 + 44 + 28 + 12 + 12 + 6 = 134$$

Percentage less than 20 minutes late

$$= \frac{32 + 44 + 28}{134}$$

$$= 77.6\% \ (3 \ \text{s.f.})$$

The shape of a distribution

In Example 1.9, you can see from the histogram that there is a very long tail to the right. This indicates that there are a few very high values.

If you superimpose a curve on a histogram or a vertical line graph, it is easier to see the general 'shape' of the distribution.

Note that the ideas of skew and skewness are not required in the examination.

Positive skew	Negative skew	Symmetrical

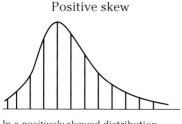 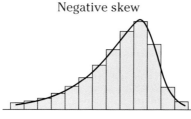 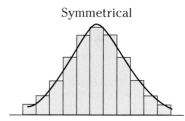

In a *positively* skewed distribution there is a long tail to the *right* (in the positive direction).

In a *negatively* skewed distribution there is a long tail to the *left* (in the negative direction).

The normal distribution (Chapter 6) is the most widely used symmetrical distribution in Statistics.

Exercise 1c

1 A researcher timed how long it took each of 38 volunteers to perform a particular task. The results are shown in the table.

Time (seconds)	$5 \leqslant t < 10$	$10 \leqslant t < 20$	$20 \leqslant t < 25$	$25 \leqslant t < 40$	$40 \leqslant t < 45$
Frequency	2	12	7	15	2

  (i) State the width of the interval $20 \leqslant t < 25$.

 (ii) Calculate the frequency density of each interval.

(iii) Draw a histogram to illustrate the data.

(iv) State the modal class.

2 On a particular day the length of stay of each car in a city car park was recorded in minutes.

Length of stay (minutes)	$0 < t < 25$	$25 \leqslant t < 60$	$60 \leqslant t < 80$	$80 \leqslant t < 150$	$150 \leqslant t < 300$
Frequency	62	70	88	280	30

Represent the data by a histogram and state the modal class.

3 The table shows the weights, to the nearest kg, of 200 girls.

Weight (kg)	41–50	51–55	56–60	61–70	71–75
Frequency	21	62	55	50	12

(i) State the boundaries of the interval 41–50.

(ii) State the width of the interval 61–70.

(iii) Draw a histogram to represent the data.

4   The daily number of visitors to a museum is recorded for a month. The results are shown in the table.

Number of visitors	30–69	70–149	150–199	200–299
Frequency	1	14	10	6

(i) State the boundaries of the interval 30–69.

(ii) Draw a histogram to represent the data.

5   These are the number of times the letter 'e' appears in each sentence in an article called 'My Kind of Day'.

　　　15   12   8   12   3   10   14   17   5   3   8   11   19
　　　　7   16   5   13   12   11   18   6   7   4   17   8   1

(i) Construct a grouped frequency table with intervals 1–2, 3–6, 7–10, 11–14, 15–19.

(ii) Draw a histogram to illustrate the data.

6   The table shows the number of letters delivered on a particular day to the individual homes in an apartment block.

Number of letters delivered, x	0–1	2–4	5–10
Number of apartments, f	12	9	3

(i) State the boundaries of the interval 0–1.

(ii) Draw a histogram to represent the data.

7   In a survey the weights of 50 apples were noted and recorded in the following table. Each weight was given to the nearest gram.

86	101	114	118	87	92	93	116	105	102
97	93	101	111	96	117	100	106	118	101
107	96	101	102	104	92	99	107	98	105
113	100	103	108	92	109	95	100	103	110
113	99	106	116	101	105	86	88	108	92

(i) Construct a frequency table, using equal intervals of width 5 grams and taking the first two intervals as 85–89, 90–94.

(ii) What is the lower boundary of the interval 90–94?

(iii) Draw a histogram to illustrate the data and state the modal class.

(iv) Draw a stem-and-leaf diagram to illustrate the data and write down the mode.

8 The histogram shows the speeds, in miles per hour, of cars passing a 30-mile-per-hour sign in Cambridge.

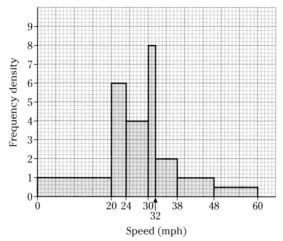

Speed (mph)

(i) Copy and complete this frequency table.

Speed (miles per hour)	$0 < x < 20$	$20 \leqslant x < 24$	$24 \leqslant x < 30$	$30 \leqslant x < 32$	$32 \leqslant x < 38$	$38 \leqslant x < 48$	$48 \leqslant x < 60$
Frequency		24					6

(ii) How many cars were observed?

9 A student recorded the length, to the nearest minute, of each lecture she attended during one particular month. She calculated the frequency density for each interval and they are shown in the table below, in which the frequencies for three of the intervals are missing.

Length of lecture (minutes)	50–53	54–55	56–59	60–67
Frequency	a	b	30	c
Frequency density	5	13	7.5	1.5

(i) Write down the boundaries of the interval 50–53 and state the width of this interval.

(ii) Calculate the values of a, b and c.

(iii) How many lectures did she attend during the month?

10 The lengths, measured to the nearest millimetre, of a random sample of pebbles taken from a particular section of a river bed are represented in the following table.

Length (mm)	1–10	11–20	21–50	51–a
Frequency	f	$2f$	150	50

(i) State the boundaries and the width of the interval 1–10.

A histogram is drawn, using a scale of 1 cm to 1 unit on the vertical axis (frequency density). The rectangle representing 1–10 has height 6 cm.

(ii) Calculate the value of f.

(iii) How many pebbles were there in the sample?

The rectangle representing 51–a has a height of 1 cm.

(iv) Calculate the value of a.

AVERAGES

An **average** value is useful when describing a set of data. This is a typical or representative value and is known as a **measure of central tendency**.

One such average is the mean.

The **mode** (page **5**) and the median (page **41**) are also averages.

The mean

The **mean** is the most commonly used average. It is calculated by dividing the sum of all the observations by the number of observations.

Consider this set of 5 numbers:

$$0.9, 1.4, 2.8, 3.1, 5.6$$

The mean is $\dfrac{0.9 + 1.4 + 2.8 + 3.1 + 5.6}{5} = \dfrac{13.8}{5} = 2.76$

In general, the **mean** of the n numbers $x_1, x_2, x_3, \dots, x_n$ is given by

$$\text{mean} = \frac{x_1 + x_2 + x_3 + \dots + x_n}{n}$$

Example 1.10

To obtain Grade A, Ben must achieve a mean mark of at least 70 in five tests. His mean mark for the first four tests is 68. What is the lowest mark Ben could achieve in the fifth test to obtain Grade A?

For the first four tests, mean mark = 68

i.e. $\dfrac{x_1 + x_2 + x_3 + x_4}{4} = 68$

so $\quad x_1 + x_2 + x_3 + x_4 = 272$

To achieve Grade A, Ben's mean mark for the five tests must be at least 70, so his **total** mark for the five tests must be $5 \times 70 = 350$.

To score a total of 350,
   his mark in fifth test must be $350 - 272 = 78$.

So, to obtain Grade A, the lowest mark Ben could achieve in the fifth test is 78.

Notation
A shorthand way of writing $x_1 + x_2 + x_3 + x_4$ is $\displaystyle\sum_{i=1}^{4} x_i$

The symbol Σ is a Greek upper case letter pronounced 'sigma' and it is used to denote 'the sum of'.

So, for $x_1 + x_2 + x_3 + \dots + x_n$ you could write $\displaystyle\sum_{i=1}^{n} x_i$

The mean is often denoted by $\bar{x}$

$\bar{x}$ is read as 'x bar'.

so $\quad \bar{x} = \dfrac{x_1 + x_2 + x_3 + \dots + x_n}{n} = \dfrac{\displaystyle\sum_{i=1}^{n} x_i}{n}$

This is very cumbersome, so reference to the subscript i is often omitted.

In shorthand notation, the **mean** of n observations is given by

$$\bar{x} = \frac{\sum x}{n}$$

Example 1.11

The members of an orchestra were asked how many instruments each could play. These are their replies.

2 5 2 4 1 1 1 2 1 3 3 2 1 2 1
1 2 4 3 2 1 2 3 1 4 2 3 1 1 2

Calculate the mean number of instruments played.

$n = 30, \sum x = 2 + 5 + 2 + \ldots + 1 + 2 = 63$

$\bar{x} = \dfrac{\sum x}{n} = \dfrac{63}{30} = 2.1$

Note that the mean is not necessarily a whole number, even if the data set consists only of whole numbers.

The mean number of instruments played is 2.1.

Mean of a frequency distribution

The data in Example 1.11 can be arranged in a frequency distribution:

Number of instruments, x	1	2	3	4	5
Frequency, f	11	10	5	3	1

The mean can be calculated from the frequency distribution as follows:

x	f	$x \times f$
1	11	11
2	10	20
3	5	15
4	3	12
5	1	5
	$\sum f = 30$	$\sum xf = 63$

$\bar{x} = \dfrac{\text{total number of instruments played}}{\text{total number of people}}$

$= \dfrac{\sum (x \times f)}{\sum f}$

$= \dfrac{63}{30}$

$= 2.1$

Do not add up the x column in a frequency distribution.

total number of people

total number of instruments played

The **mean** of data in a **frequency distribution** is given by

$\bar{x} = \dfrac{\sum xf}{\sum f}$

Note that $\sum xf$ means $\sum (x \times f)$.

What happens when the data have been grouped?

When you have a **grouped frequency distribution**, the actual data values are no longer available, so you can only make an **estimate** of the mean.

You need a representative value for each interval, and this is taken to be the **mid-interval value** or mid-point of the interval.

Mid-interval value $= \frac{1}{2}$(lower class boundary + upper class boundary)

The mid-interval value is the **mean** of the class boundaries.

Example 1.12

In a spot check, the speeds of 120 vehicles on a particular stretch of road through a village were noted. The results are shown in the table.

Speed x km/h	21–25	26–30	31–35	36–45	46–60
Frequency f	22	48	25	16	9

Estimate the mean speed of these vehicles.

Work out the mid-interval value for the first interval 21–25.

The lower boundary = 20.5 and the upper boundary = 25.5, so
mid-interval value = $\frac{1}{2}(20.5 + 25.5) = 23$.

You now assume that all the values in the first interval are 23 and use this value for x. Find all the mid-interval values and form a table.

Speed (km/h)	Mid-interval value, x	f	$x \times f$
21–25	23	22	506
26–30	28	48	1344
31–35	33	25	825
36–45	40.5	16	648
46–60	53	9	477
		$\sum f = 120$	$\sum xf = 3800$

Do not add this column.

$$\bar{x} = \frac{\sum xf}{\sum f}$$
$$= \frac{3800}{120}$$
$$= 31.7 \text{ km/h (3 s.f.)}$$

Example 1.13

The diagram shows a histogram of the distribution of the weights of 50 first-year students at a particular university. All the rectangles have been drawn, but the vertical scale is missing.

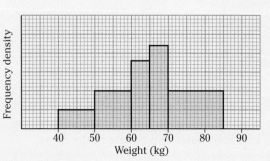

(i) Compile a grouped frequency table for the distribution.

(ii) Use the values in your frequency table to find an estimate of the mean weight of these students.

Let the height of one small square on the vertical axis be h.

Total area of the histogram

$= 10 \times 5h + 10 \times 10h + 5 \times 18h + 5 \times 22h + 15 \times 10h$

$= 50h + 100h + 90h + 110h + 150h$

$= 500h$

> The area of each rectangle gives the frequency in that interval, so the total area gives the total frequency.

Total area = total frequency,

so $\quad\quad 500h = 50$

$\quad\quad\quad\quad h = 0.1$

The frequency in the first interval is $50 \times 0.1 = 5$.

Calculate all the frequencies and write them in the table.   Check that the total frequency is 50.

The frequencies are   5, 10, 9, 11, 15.

Calculate the mid-interval value of each interval.

Weight (kg)	Mid-interval value, x	f	$x \times f$
$40 \leqslant x < 50$	45	5	225
$50 \leqslant x < 60$	55	10	550
$60 \leqslant x < 65$	62.5	9	562.5
$65 \leqslant x < 70$	67.5	11	742.5
$70 \leqslant x < 85$	77.5	15	1162.5
		$\sum f = 50$	$\sum xf = 3242.5$

$\bar{x} = \dfrac{\sum xf}{\sum f}$

$\quad = \dfrac{3242.5}{50}$

$\quad = 64.85 \, \text{kg}$

Example 1.14

The following table shows the results of a survey to find the average daily time, in hours, that people spent watching television.

Time spent per day (t hours)	$0 \leqslant t < 1$	$1 \leqslant t < 2$	$2 \leqslant t < 4$	$4 \leqslant t < 8$
Number of people	10	18	f	4

An estimate of the mean time was calculated to be 2 hours.

Form an equation involving f and hence find the number of people in the survey.

Find the mid-interval values.

The mid-interval values are 0.5, 1.5, 3, 6

Estimate of total number of hours

$$= (0.5 \times 10) + (1.5 \times 18) + (3 \times f) + (6 \times 4)$$

$$= 56 + 3f$$

Total number of people

$$= 10 + 18 + f + 4$$

$$= 32 + f$$

$$\text{Mean} = \frac{\text{Total number of hours}}{\text{Total number of people}}$$

$$= \frac{56 + 3f}{32 + f}$$

so $\dfrac{56 + 3f}{32 + f} = 2$

You are given that the mean is 2.

$$56 + 3f = 2(32 + f)$$

Solve the equation.

$$56 + 3f = 64 + 2f$$

$$f = 8$$

Number of people in survey $= 32 + 8 = 40$.

Using a calculator in statistical mode

When finding the mean you can use your calculator in **computation mode** to calculate the total and also do the division. However, you will find it very useful to be able to find the mean *directly* using the **statistical mode** on your calculator.

An outline is shown below, but you may need to consult your calculator manual for the details.

Set the statistical mode, sometimes written SD or STAT.

Clear the statistical registers (memories).

Input the data. It should be possible for you to input either individual values or values from a frequency distribution in the order x then f.

You should then have access to the following:

$\boxed{\bar{x}}$ (the mean), $\boxed{n}$ (the number of values), $\boxed{\textstyle\sum x}$ (the sum of the values).

If you are inputting data from a frequency table,

$\sum f$ is given by $\boxed{n}$ and $\sum xf$ is given by $\boxed{\textstyle\sum x}$.

Practise by using the data in Examples 1.11, 1.12 and 1.13 so that you are proficient in the use of the calculator.

Exercise 1d

1 Find the mean of each of these sets of numbers without using the statistical mode on your calculator. Then check using the statistical mode.

(i) 5, 6, 6, 8, 8, 9, 11, 13, 14, 17

(ii) 148, 153, 156, 157, 160

(iii) 44, 47, 48, 51, 52, 54, 55, 56

(iv) 1769, 1771, 1772, 1775, 1778, 1781, 1784

(v) 0.85, 0.88, 0.89, 0.93, 0.94, 0.96

(vi)

x	1	2	3	4	5	6	7
f	4	5	8	10	17	5	1

(vii)

x	27	28	29	30	31	32
f	30	43	51	49	42	35

2 Find the mean of the data represented in this stem-and-leaf diagram.

```
1 | 2 8
2 | 1 4 4 6
3 | 0 0 2 3 5 5 5
4 | 2 3 6 7
5 | 3 6 9
```

Key: 1 | 2 means 12 grams

3 The age distribution of the population of a small village is recorded in the table.

Age	Number of people
$0 < x < 15$	54
$15 \leqslant x < 30$	78
$30 \leqslant x < 50$	120
$50 \leqslant x < 70$	88
$70 \leqslant x < 100$	60

Estimate the mean age.

4 A stock check was carried out on the number of books on a section of shelves in a library.

Number of books on shelf	Number of shelves
31–35	4
36–40	6
41–45	10
46–50	13
51–55	5
56–60	2

(i) State the mid-interval value of the first interval.

(ii) Estimate the mean number of books on a shelf.

5   The height, to the nearest metre, was recorded of each of the 215 birch trees in a wood. The heights are summarised in the frequency table.

Height (m)	5–9	10–12	13–15	16–18	19–28
Frequency	30	43	51	49	42

Estimate the mean height of the birch trees in the wood.

6   In a busy office, the duration of time the telephone rang before it was answered was noted for 105 calls. The times were recorded in seconds, to the nearest second. The results are summarised in the table.

Time (to the nearest second)	Number of calls
10–19	20
20–24	20
25–29	15
30–31	14
32–34	16
35–39	10
40–59	10

(i) Estimate the mean time before a telephone call was answered.

(ii) What is the modal class?

7   The frequency distribution shows the marks achieved in a test by 100 students.

Test mark	30–39	40–49	50–59	60–69	70–79	80–99
Frequency	10	14	26	20	18	12

Estimate the mean mark.

8   The mean of this frequency distribution is 3.66.

x	1	2	3	4	5	6
f	3	9	a	11	8	7

Find a.

9   The lengths, in metres, of the gardens of the houses in a particular street are represented in the following table.

Length (m)	$10 \leqslant x < 16$	$16 \leqslant x < 20$	$20 \leqslant x < 40$	$40 \leqslant x < 50$
Frequency	4	12	f	8

An estimate of the mean length is 27.7 m.

(i) Find f. 　　　　　　　　　　　　　　(ii) How many houses are there in the street?

10   A bag contained five balls each bearing a different number: 1, 2, 3, 4 or 5. A ball was drawn from the bag, its number was noted, and then it was put back in the bag. This was done 50 times in all and the table below shows the resulting frequency distribution.

Number	1	2	3	4	5
Frequency	x	11	y	8	9

The mean of this frequency distribution is 2.7. Find x and y.

11 The mean of ten numbers is 8. When an eleventh number is included, the mean is 9. What is the value of the eleventh number?

12 The mean of a list of 8 numbers is 15. When two more numbers, x and $2x$, are added to the list, the mean is 13.2. Find x.

13 The mean of four numbers is 5, and the mean of three different numbers is 12. What is the mean of the seven numbers?

14 The mean of n numbers is 5. If the number 13 is included with the n numbers, the new mean is 6. Find n.

VARIABILITY OF DATA

Consider these three sets of numbers, each with a mean of 7:

(a) $\boxed{7, 7, 7, 7, 7}$ (b) $\boxed{4, 6, 6.5, 7.2, 11.3}$ (c) $\boxed{-193, -46, 28, 69, 177}$

Although the means are the same, the spread of each is set is different. There is no variability in set (a), but the numbers in set (c) are much more spread out than those in set (b).

There are various ways of measuring the **variability** or **spread** of a distribution. Two of these, the range and the standard deviation, are described in this section.

Range

The **range** is based entirely on the extreme values of the distribution and gives a quick snapshot of the overall spread of the data.

Range = highest value − lowest value

Note that the **interquartile range**, which is based on particular observations **within** the data, is described on page **42**.

In the sets of numbers above:

Set (a) has a range of $7 - 7 = 0$

Set (b) has a range of $11.3 - 4 = 7.3$

Set (c) has a range of $177 - (-193) = 370$

Standard deviation

Standard deviation is a very useful measure of spread and it is particularly important in later work. It gives a measure of the spread of the data in relation to the mean, $\bar{x}$, of the distribution. It is calculated using all the values in the distribution, as follows:

For each value x, calculate how far it is from the mean by finding $(x - \bar{x})$. This is the <u>deviation</u> from the mean.

Note that $(x - \bar{x})$ could be positive or negative, depending on whether x is above the mean or below the mean.

However, $(x - \bar{x})^2 \geqslant 0$.

Now square the deviation to give $(x - \bar{x})^2$.

Find the sum of these squares, written $\sum(x - \bar{x})^2$.

Then divide by n, the number of values, to give the **variance** $\dfrac{\sum(x - \bar{x})^2}{n}$.

Finally take the positive square root to give the **standard deviation**, where

$$\text{standard deviation} = \sqrt{\frac{\sum(x - \bar{x})^2}{n}}.$$

The **standard deviation** of a set of n numbers with mean $\bar{x}$ is given by

Remember that standard deviation is **never** negative.

$$\text{s.d.} = \sqrt{\frac{\sum(x - \bar{x})^2}{n}}$$

These are the standard deviations for the three sets of numbers, where $\bar{x} = 7$ for each set:

Set (a) 7, 7, 7, 7, 7

  Since $(x - \bar{x}) = 7 - 7 = 0$ for *every* value, there is <u>no deviation</u> from the mean.
  So s.d. = 0.

Set (b) 4, 6, 6.5, 7.2, 11.3

  $\sum(x - \bar{x})^2 = (4 - 7)^2 + (6 - 7)^2 + (6.5 - 7)^2 + (7.2 - 7)^2 + (11.3 - 7)^2 = 28.78$

  So s.d. $= \sqrt{\dfrac{\sum(x - \bar{x})^2}{n}} = \sqrt{\dfrac{28.78}{5}} = 2.4$ (1 d.p.).

Set (c) −193, −46, 28, 69, 177

  $\sum(x - \bar{x})^2 = (-193 - 7)^2 + (-46 - 7)^2 + (28 - 7)^2 + (69 - 7)^2 + (177 - 7)^2 = 75\,994$

  So s.d. $= \sqrt{\dfrac{\sum(x - \bar{x})^2}{n}} = \sqrt{\dfrac{75\,994}{5}} = 123.3$ (1 d.p.).

Set (c) has a much higher standard deviation than set (b), confirming that it is much more spread about the mean than set (b).

Standard deviation is very useful for comparing distributions: the **lower** the standard deviation, the **less variation** there is and the **more consistent** the data are.

The 'definition version' for standard deviation (given above) can be difficult to use, especially when $\bar{x}$ is not an integer, so it is more usual to use an alternative form. This is sometimes referred to as the 'calculation' version.

'Calculation' version of formula for standard deviation:

$$\text{s.d.} = \sqrt{\dfrac{\sum x^2}{n} - \bar{x}^2} \qquad \text{where } \bar{x} = \dfrac{\sum x}{n}$$

Example 1.15

The mean of the numbers 2, 3, 5, 6, 8 is 4.8. Calculate the standard deviation.

Method 1: Definition version		
x	$x - 4.8$	$(x - 4.8)^2$
2	−2.8	7.84
3	−1.8	3.24
5	0.2	0.04
6	1.2	1.44
8	3.2	10.24
		$\sum(x - \bar{x})^2 = 22.8$

$$\text{s.d.} = \sqrt{\dfrac{\sum(x - \bar{x})^2}{n}}$$
$$= \sqrt{\dfrac{22.8}{5}}$$
$$= 2.14 \text{ (3 s.f.)}$$

Method 2: Calculation version	
x	x^2
2	4
3	9
5	25
6	36
8	64
	$\sum x^2 = 138$

$$\text{s.d.} = \sqrt{\dfrac{\sum x^2}{n} - \bar{x}^2}$$
$$= \sqrt{\dfrac{138}{5} - 4.8^2}$$
$$= 2.14 \text{ (3 s.f.)}$$

You are strongly advised to use the 'calculation' version, rather than the 'definition' version in your calculations.

Variance

The **variance** is the square of the standard deviation, where

$$\text{variance} = \frac{\sum(x - \bar{x})^2}{n} = \frac{\sum x^2}{n} - \bar{x}^2$$

Remember
$$\text{variance} = (\text{standard deviation})^2$$
$$\text{standard deviation} = \sqrt{\text{variance}}$$

Derivation of calculation version of standard deviation

It is interesting to see how the calculation version for standard deviation is derived from the definition version. Note, however, that this derivation is not needed in the examination.

Start by considering the variance:

$$\text{variance} = \frac{\sum(x - \bar{x})^2}{n}$$

$$= \frac{1}{n}\sum(x - \bar{x})^2$$

$$= \frac{1}{n}\sum(x^2 - 2x\bar{x} + \bar{x}^2)$$

$$= \frac{1}{n}\left(\sum x^2 - 2\bar{x}\sum x + \sum \bar{x}^2\right)$$

$$= \frac{\sum x^2}{n} - 2\bar{x}\frac{\sum x}{n} + \frac{n\bar{x}^2}{n}$$

$$= \frac{\sum x^2}{n} - 2\bar{x}(\bar{x}) + \bar{x}^2 \qquad \text{since } \frac{\sum x}{n} = \bar{x}$$

$$= \frac{\sum x^2}{n} - \bar{x}^2$$

The variance can be thought of as 'the mean of the squares minus the square of the mean'.

Now take the positive square root to give the standard deviation.

$$\text{s.d.} = \sqrt{\frac{\sum x^2}{n} - \bar{x}^2}$$

Standard deviation of a frequency distribution

When data are in a **frequency distribution**, the formula for the standard deviation is as follows:

Definition version

$$\text{s.d.} = \sqrt{\frac{\sum(x - \bar{x})^2 f}{\sum f}}$$

Calculation version

$$\text{s.d.} = \sqrt{\frac{\sum x^2 f}{\sum f} - \bar{x}^2} \qquad \text{where } \bar{x} = \frac{\sum xf}{\sum f}$$

Example 1.16

The distribution shows the number of children in 20 families. The mean number of children in a family is 2.9. Calculate

(i) the range

(ii) the standard deviation.

Number of children in family, x	1	2	3	4	5
Number of families, f	3	4	8	2	3

(i)   Range = highest value − lowest value
      = 5 − 1 = 4

(ii) *Use the calculation version.*

Do NOT square f

x	x^2	f	$x^2 \times f$
1	1	3	3
2	4	4	16
3	9	8	72
4	16	2	32
5	25	3	75
		$\sum f = 20$	$\sum x^2 f = 198$

Do **not** add the x column
or the x^2 column.

$$\text{s.d.} = \sqrt{\frac{\sum x^2 f}{\sum f} - \bar{x}^2}$$
$$= \sqrt{\frac{198}{20} - 2.9^2}$$
$$= \sqrt{1.49}$$
$$= 1.22 \text{ (3 s.f.)}$$

When the data are **grouped**, estimate the standard deviation by using the mid-interval value for x.

Example 1.17

An online test was taken by 115 students. The time spent on each question was recorded by the computer. The following table shows the time taken, in minutes, on the final question.

Time (mins)	$1 \leqslant x < 2$	$2 \leqslant x < 3$	$3 \leqslant x < 5$	$5 \leqslant x < 10$
Frequency	16	32	42	25

Calculate estimates of the mean and standard deviation of the time spent on the final question.

Time (mins)	Mid-interval value, x	f	$x \times f$	$x^2 \times f$
$1 \leqslant x < 2$	1.5	16	24	36
$2 \leqslant x < 3$	2.5	32	80	200
$3 \leqslant x < 5$	4	42	168	672
$5 \leqslant x < 10$	7.5	25	187.5	1406.25
		$\sum f = 115$	$\sum xf = 459.5$	$\sum x^2 f = 2314.25$

$$\bar{x} = \frac{\sum xf}{\sum f} = 3.995\ldots = 4.00 \text{ mins (3 s.f.)}$$

$$\text{s.d.} = \sqrt{\frac{\sum x^2 f}{\sum f} - \bar{x}^2}$$

$$= \sqrt{\frac{2314.25}{115} - (3.995\ldots)^2}$$

$$= 2.039\ldots$$

$$= 2.04 \text{ mins (3 s.f.)}$$

Using the calculator to find the standard deviation

The standard deviation of a set of data can be found directly on the calculator in statistical mode. The numbers are entered in the same way as when finding the mean (see page 24)

You should then have access to the standard deviation which is often labelled $\boxed{_x\sigma_n}$.

Note that the key labelled $\boxed{_x\sigma_{n-1}}$ gives an unbiased estimate of a population standard deviation based on a sample. It will not be required until module S2.

You can also get $\boxed{\sum x^2}$. Remember that when inputting the data from a frequency table, this gives $\sum x^2 f$.

Practise by using the data in Examples 1.15, 1.16 and 1.17.

Examination advice

If you use the statistical functions in the examination and write down only the answer, you will get full marks if the answer is correct and no marks if it is incorrect. So you are advised to do every calculation twice to check. Also look at the size of your answer and decide whether it is sensible.

You could use the statistical functions but also write down the totals from your calculator and show them substituted into the formula.

If you are not using the statistical functions, show a table with totals and substitute them into the appropriate formula.

The following are given on the formulae sheet in the examination:

For raw data:

$$\bar{x} = \frac{\sum x}{n} \qquad\qquad \text{standard deviation} = \sqrt{\frac{\sum (x - \bar{x})^2}{n}} = \sqrt{\frac{\sum x^2}{n} - \bar{x}^2}$$

For data in a frequency table:

Use the calculation versions.

$$\bar{x} = \frac{\sum xf}{\sum f} \qquad\qquad \text{standard deviation} = \sqrt{\frac{\sum (x - \bar{x})^2 f}{\sum f}} = \sqrt{\frac{\sum x^2 f}{\sum f} - \bar{x}^2}$$

Exercise 1e

1 *Do not use the statistical functions on your calculator for part (i) of this question.*

   (i) For each of the following sets of numbers, calculate the mean and the standard deviation.

   (a) 2, 4, 5, 6, 8   (b) 6, 8, 9, 11

   (c) 11, 14, 17, 23, 29   (d) 5, 13, 7, 9, 16, 15

   (e) 4.6, 2.7, 3.1, 0.5, 6.2   (f) 200, 203, 206, 207, 209

   (ii) Now check your answers using your calculator in statistical mode.

2 The stem-and-leaf diagram shows the times taken by 10 people to run a race.

Stem	Leaf
12	8
13	4 6 6 8
14	7 9
15	0 7
16	4

Key: 12 | 8 means 12.8 seconds

Calculate the mean and standard deviation of the times.

3 *Do this question without using the statistical functions on your calculator and then check using the statistical mode.*

x	10	11	12	13
f	3	7	11	2

Calculate the mean and standard deviation.

4 The score for a round of golf for each of 50 club members was recorded and the results are summarised in the frequency table.

Score, x	66	67	68	69	70	71	72	73
Frequency, f	2	5	10	12	9	6	4	2

Calculate the mean score and the standard deviation.

5 A vertical line graph for a set of data is shown below.

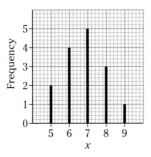

Calculate the mean and standard deviation of the data.

6 A group of 20 people play a game. The table shows the frequencies of their scores.

Score	1	2	4	x
Frequency	2	5	7	6

The mean score is 5.

(i) Find x.

(ii) Calculate the standard deviation.

7 Becky plays a computer game where she fires at a target. Her score is 1 if she hits the target and 0 if she misses it.

She has 30 attempts and hits the target 18 times.

(i) Find her mean score for the 30 attempts.

(ii) Find the variance of her scores for the 30 attempts.

8 The floor areas, $x\,\mathrm{m}^2$, of 105 houses in a housing development are as follows:

Floor area ($x\,\mathrm{m}^2$)	$50 \leqslant x < 150$	$150 \leqslant x < 200$	$200 \leqslant x < 250$	$250 \leqslant x < 350$	$350 \leqslant x < 500$
Frequency	20	29	36	15	5

Calculate estimates of the mean and standard deviation of the floor areas.

9   The scores of 60 candidates in an intelligence test are shown in the table.

Score	100–106	107–113	114–120	121–127	128–134
Frequency	8	13	24	11	4

Calculate estimates of the mean score and the standard deviation.

10   A survey is carried out to determine the numbers of pupils in various age groups attending nurseries, schools and colleges in a certain region. The results are summarised in the table.

Age, x years	$3 \leqslant x < 5$	$5 \leqslant x < 7$	$7 \leqslant x < 11$	$11 \leqslant x < 16$	$16 \leqslant x < 18$
Frequency	58	127	350	567	398

Calculate estimates of the mean age and the standard deviation.

11   A school librarian noted the number of books borrowed from the library during a month by the pupils in a particular year group. The results are summarised in the table.

Number of books	0–2	3–5	6–10	11–15
Number of pupils	21	54	22	13

   (i) State the class boundaries of the first group.

   (ii) Calculate estimates of the mean and standard deviation.

12   At a bird observatory, migrating willow warblers are caught, measured and ringed before being released. The histogram illustrates the lengths, in millimetres, of the willow warblers caught during one migration season.

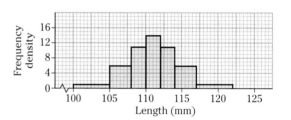

   (i) Copy and complete the frequency table:

Length (mm)	$100 \leqslant x < 105$	$105 \leqslant x < 108$	$108 \leqslant x < 110$	$110 \leqslant x < 112$	$112 \leqslant x < 114$	$114 \leqslant x < 117$	$117 \leqslant x < 122$
Frequency						18	

   (ii) State briefly how it may be deduced from the histogram (without calculating) that an estimate of the mean length is 111 mm and explain why this value may not be the true mean length of the willow warblers caught.

   (iii) Estimate the standard deviation of the lengths.

13   Thirty-one people completed a jigsaw in the following times (x minutes).

11 53 72 48 48   49 39 87 73 23   120 24 61 36   66 67
86 79 65 47 36 133 78 81 70 75   53 42 42 72 144

   (i) Calculate the mean $\bar{x}$.

   (ii) Calculate the standard deviation, s.d.

The following rule can be used to identify unusually high or low values, called outliers.

'An outlier is less than $\bar{x} - 2 \times$ (s.d.) or greater than $\bar{x} + 2 \times$ (s.d.)'

   (iii) Use this rule to identify any outliers.

COMBINING SETS OF DATA

When you are not given the actual data but instead are given totals such as $\sum x$ and $\sum x^2$, you will need to substitute them into the formulae for the mean and the standard deviation. This is particularly useful when combining sets of data.

Example 1.18

The ages, x years, of 18 people attending an evening class are summarised by the following totals:

$$\sum x = 745, \sum x^2 = 33\,951.$$

(i) Calculate the mean and standard deviation of the ages of this group of people.

(ii) One person leaves the group and the mean age of the remaining 17 people is exactly 41 years. Find the age of the person who left and the standard deviation of the ages of the remaining 17 people.

Cambridge Paper 6 Q4 N04

(i) $\bar{x} = \dfrac{\sum x}{n} = \dfrac{745}{18} = 41.388... = 41.4$ years (3 s.f.)

$\text{s.d.} = \sqrt{\dfrac{\sum x^2}{n} - \bar{x}^2} = \sqrt{\dfrac{33\,951}{18} - (41.38...)^2} = 13.157... = 13.2$ years (3 s.f.)

(ii) For 17 people:

Mean $= 41$, so $\sum x = 41 \times 17 = 697$

Previous total $= 745$, so age of leaver $= 745 - 697 = 48$

New $\sum x^2 = 33\,951 - 48^2 = 31\,647$

New standard deviation

$$= \sqrt{\dfrac{31\,647}{17} - 41^2} = 13.43... = 13.4 \text{ years (3 s.f.)}$$

In general, for two sets of data, x and y,

$$\text{mean} = \frac{\sum x + \sum y}{n_1 + n_2} \qquad\qquad \text{standard deviation} = \sqrt{\frac{\sum x^2 + \sum y^2}{n_1 + n_2} - (\text{mean})^2}$$

Example 1.19

The following table shows the mean and standard deviation of the heights of 20 boys and 30 girls.

	Mean	Standard deviation
Boys	160 cm	4 cm
Girls	155 cm	3.5 cm

Find the mean and standard deviation of the heights of the 50 children.

Boys:

$n_1 = 20$ and $\bar{x} = 160$, so $\sum x = 160 \times 20 = 3200$

You are given that the standard deviation is 4, where s.d. $= \sqrt{\dfrac{\sum x^2}{n} - \bar{x}^2}$

So $\sqrt{\dfrac{\sum x^2}{20} - 160^2} = 4$

Squaring both sides:

$$\dfrac{\sum x^2}{20} - 160^2 = 4^2 \qquad\qquad \text{This is the variance.}$$

$$\sum x^2 = 20(4^2 + 160^2) = 512\,320$$

Girls:

$n_2 = 30$ and $\bar{y} = 155$, so $\sum y = 155 \times 30 = 4650$

s.d. $= 3.5$, where s.d. $= \sqrt{\dfrac{\sum y^2}{n} - \bar{y}^2}$

So $\dfrac{\sum y^2}{30} - 155^2 = 3.5^2$

$$\sum y^2 = 30(3.5^2 + 155^2) = 721\,117.5$$

All 50 children:

New mean $= \dfrac{\sum x + \sum y}{n_1 + n_2} = \dfrac{3200 + 4650}{50} = 157$

New standard deviation $= \sqrt{\dfrac{\sum x^2 + \sum y^2}{n_1 + n_2} - (\text{mean})^2}$

$$= \sqrt{\dfrac{512\,320 + 721\,117.5}{50} - 157^2}$$

$$= 4.444\ldots$$

$$= 4.44 \ (3 \text{ s.f.})$$

'Coding' data

Example 1.20

Sweets are packed into bags with a nominal weight of 75 grams. Ten bags are picked at random from the production line and weighed. Their weights, in grams, are

76.0, 74.2, 75.1, 73.7, 72.0, 74.3, 75.4, 74.0, 73.1, 72.8

(i)   Use your calculator to find the mean and the standard deviation.

(ii) It is later discovered that the scales were reading 3.2 grams below the correct weight.

    (a)   What was the correct mean weight of the 10 bags?

    (b)   What was the correct standard deviation of the 10 bags?

(i) Using a calculator, you should find that
Mean = 74.06 grams, standard deviation = 1.166... = 1.17 grams (3 s.f.)

(ii) The correct readings are
79.2, 77.4, 78.3, 76.9, 75.2, 77.5, 78.6, 77.2, 76.3, 76.0

(a) Correct mean = 77.26 grams

(b) Correct standard deviation = 1.166... = 1.17 grams (3 s.f.)

Notice that when each weight is increased by 3.2 grams,
– the mean weight is increased by 3.2 grams,
– the standard deviation is unaltered.

When the two sets of data are shown on the same diagram, it is easy to see why the standard deviation is unaltered as the spread of the original data about the original mean is the same as the spread of the new data about the new mean.

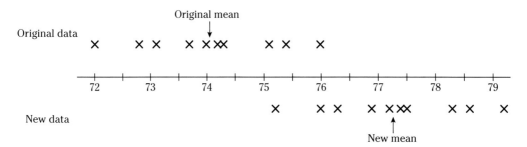

In general, if each data value is increased by a constant a

• the mean is increased by a

• the standard deviation is unaltered.

This is particularly useful when finding the mean and standard deviation using $\sum(x - a)$ and $\sum(x - a)^2$, where a is a constant.

The constant a is sometimes called the 'working mean'.

Finding the mean and standard deviation using $\sum(x - a)$ and $\sum(x - a)^2$

To find the mean $\bar{x}$

• Find the mean of $(x - a)$
• Now add a

To find the standard deviation

• Find the standard deviation of $(x - a)$
• This is the same as the standard deviation of x.

Example 1.21

The time taken, x minutes, by Katy to do the Sudoku puzzle in a certain newspaper was observed on 20 occasions. The results are summarised below.

$$\sum(x - 30) = -50 \qquad \sum(x - 30)^2 = 562$$

Find the mean and standard deviation of the time taken by Katy to solve the Sudoku puzzle.

Mean

First find the mean of (x − 30).

$$\frac{\sum(x - 30)}{20} = \frac{-50}{20} = -2.5$$

Now add 30 to find the mean of x.

So $\bar{x} = -2.5 + 30 = 27.5$

The mean time taken to complete the Sudoku puzzle is 27.5 minutes.

Standard deviation

First find the s.d. of (x − 30).

$$\text{s.d. of } (x - 30) = \sqrt{\frac{\sum(x - 30)^2}{20} - (-2.5)^2}$$

The mean of (x − 30) is −2.5.

$$= \sqrt{\frac{562}{20} - 6.25}$$

$$= 4.674\ldots$$

This is the same as the s.d. of x.

So s.d. of $x = 4.674\ldots$

The standard deviation of the times taken to solve the Sudoku puzzle is 4.67 minutes (3 s.f.).

Note:
The value of $\sum(x - a)$ gives you the following information about the mean.

- When $\sum(x - a) < 0$, the mean is less than a (as in the example above).
- When $\sum(x - a) = 0$, the mean **is** a.
- When $\sum(x - a) > 0$, the mean is greater than a.

Using formulae
You could use the following formulae:

$$\bar{x} = \frac{\sum(x - a)}{n} + a$$

mean of $(x - a)$

$$\text{s.d. of } x = \sqrt{\frac{\sum(x - a)^2}{n} - \left(\frac{\sum(x - a)}{n}\right)^2}$$

mean of $(x - a)$

However, these are NOT given in the examination, so you are advised to remember the method, rather than risk misquoting them.

Example 1.22

A summary of 24 observations of x gave the following information:

$$\sum(x - a) = -73.2 \quad \text{and} \quad \sum(x - a)^2 = 2115$$

The mean of these values of x is 8.95.

(i) Find the value of the constant a.

(ii) Find the standard deviation of these values of x. 　　　Cambridge Paper 6 Q1 N07

(i) $n = 24$

mean of x = mean of $(x - a) + a$

So, $8.95 = \dfrac{\sum(x - a)}{24} + a$

$8.95 = \dfrac{-73.2}{24} + a$

$8.95 = -3.05 + a$

$a = 8.95 + 3.05 = 12$

(ii) s.d. of $(x - a) = \sqrt{\dfrac{\sum(x - a)^2}{24} - (-3.05)^2}$ ⟵ From Part (i), the mean of $(x - a)$ is -3.05.

$= \sqrt{\dfrac{2115}{24} - 3.05^2}$

$= 8.8782...$

s.d. of x = s.d. of $(x - a)$

$= 8.88$ (3 s.f.)

Example 1.23

Delip measured the speeds, x km per hour, of 70 cars on a road where the speed limit is 60 km per hour. His results are summarised by $\sum(x - 60) = 245$.

(i) Calculate the mean speed of these cars.

His friend Sachim used values of $(x - 50)$ to calculate the mean.

(ii) Find $\sum(x - 50)$.

(iii) The standard deviation of the speeds is 10.6 km per hour. Calculate $\sum(x - 50)^2$.

Cambridge Paper 63 Q4 N10

(i) $\bar{x}$ = mean of $(x - 60) + 60$

$= \dfrac{\sum(x - 60)}{70} + 60$

$= \dfrac{245}{70} + 60$

$= 3.5 + 60$

$= 63.5$

As $\sum(x - 60) > 0$, you know that $\bar{x}$ is greater than 60.

Use this fact to check that your answer is reasonable.

(ii) $\bar{x}$ = mean of $(x - 50) + 50$

so $63.5 = \dfrac{\sum(x - 50)}{70} + 50$

$\dfrac{\sum(x - 50)}{70} = 63.5 - 50 = 13.5$

so $\sum(x - 50) = 13.5 \times 70 = 945$

Were you expecting $\sum(x - 50)$ to be positive or negative?

(iii)

$$\text{s.d. of } x = \text{s.d. of } (x - 50)$$

so

$$10.6 = \sqrt{\frac{\sum(x-50)^2}{70} - 13.5^2}$$

From Part (i), mean of $(x - 50) = 13.5$

$$10.6^2 = \frac{\sum(x-50)^2}{70} - 13.5^2$$

Square both sides. Note that it is easier to work in variances here than standard deviations.

$$\frac{\sum(x-50)^2}{70} = 10.6^2 + 13.5^2$$

$$= 294.61$$

So

$$\sum(x-50)^2 = 294.61 \times 70$$

$$= 20\,622.7$$

Exercise 1f

To answer these questions you need to be familiar with the formulae for the mean and the standard deviation.

1 Cartons of orange juice are advertised as containing 1 litre. A random sample of 100 cartons gave the following results for the volume, x litres.

$$\sum x = 101.4 \text{ and } \sum x^2 = 102.83$$

Calculate the mean and standard deviation of the volume of orange juice in these 100 cartons.

2 For a set of 10 numbers, $\sum x = 290$ and $\sum x^2 = 8469$.

Find (i) the mean, (ii) the standard deviation, (iii) the variance.

3 For a particular set of data,

$n = 100, \sum x = 584$ and $\sum x^2 = 23\,781$.

Find

(i) the mean, $\bar{x}$

(ii) the variance.

4 For a set of 9 numbers, $\sum(x - \bar{x})^2 = 234$. Find the standard deviation of the numbers.

5 For a set of 12 numbers, it is given that

$\sum(x - \bar{x})^2 = 60$, where $\bar{x}$ is the mean.

(i) Find the standard deviation.

It is also given that $\sum x^2 = 285$.

(ii) Find $\bar{x}$.

6 From the information given about each of the following sets of data, work out the missing values in the table:

	n	$\sum x$	$\sum x^2$	$\bar{x}$	s.d.
(i)	63	7623	924 800		
(ii)		152.6		10.9	1.7
(iii)	52		57 300	33	
(iv)	18			57	4

7 A machine cuts lengths of wood. A sample of 20 rods gave the following results for the length, x cm.

$$\sum xf = 997 \qquad \sum x^2 f = 49\,711$$

(i) Find the mean length of the 20 rods.

(ii) Find the variance of the lengths of the 20 rods.

8 For a particular set of observations,

$$\sum f = 20, \sum x^2 f = 16\,143, \sum xf = 563.$$

Calculate the standard deviation.

9 The mean of the numbers 3, 6, 7, a, 14, is 8. Find the standard deviation of the set of numbers.

10 (i) Calculate the mean and the standard deviation of the four numbers 2, 3, 6, 9.

(ii) When two numbers, a and b, are added to this set of four numbers, the mean is increased by 1 and the **variance** is increased by 2.5. Find a and b.

11 The numbers $a, b, 8, 5, 7$ have mean 6 and variance 2. Find the values of a and b, if $a < b$.

12 A test is taken by 30 students. Their scores, x, have a mean of 60 and a standard deviation of 20.

   (i) Find $\sum x$ and show that $\sum x^2 = 120\,000$.

   Another 20 students take the test. Their scores, y, are such that $\sum y = 1400$ and $\sum y^2 = 100\,000$.

   (ii) Show that the mean score of the combined group of 50 students is 64.

   (iii) Calculate the standard deviation of the scores of the 50 pupils.

13 The number of errors, x, on each of 200 pages of typescript was monitored. The results were summarised as follows:

$$\sum x = 920 \qquad \sum x^2 = 5032$$

   (i) Calculate the mean and standard deviation of the number of errors on a page.

   A further 50 pages were monitored and it was found that, for these pages, the mean was 4.4 errors and the standard deviation was 2.2 errors.

   (ii) Find the mean and standard deviation of the number of errors per page for the 250 pages.

14 The manager of a car showroom monitored the numbers of cars sold during two successive five-day periods.

   During the first five days the numbers of cars sold per day had mean 1.8 and standard deviation 0.6. During the next five days the numbers of cars sold per day had mean 2.8 and standard deviation 0.81.

Find the mean and standard deviation of the numbers of cars sold per day during the full ten days.

15 For a particular set of data,

$n = 100, \sum(x - 50) = 123.5$ and $\sum(x - 50)^2 = 238.4$.

Find the mean and standard deviation of x.

16 Salt is packed in bags which the manufacturer claims contain 25 kg. Eighty bags are examined and the weight, x kg, of each bag is found. The results are summarised as follows.

$$\sum(x - 25) = 27.2 \quad \text{and} \quad \sum(x - 25)^2 = 85.1$$

   (i) Without doing any calculations, state whether the mean weight of the 80 bags is less than 25 kg, equal to 25 kg or more than 25 kg.

   (ii) Calculate the mean weight.

   (iii) Find the standard deviation of the weights of the 80 bags.

17 A summary of 15 observations of x gave the following information:

$$\sum(x - 2.5) = 200$$

Find (i) the mean $\bar{x}$, (ii) $\sum(x - \bar{x})$.

18 It is known that, for 100 observations of x, the mean $\bar{x} = 25$.

Find

   (i) $\sum x$ (ii) $\sum(x - 20)$ (iii) $\sum(x - 27)$.

Given also that the standard deviation of x is 3, find

   (iv) $\sum x^2$ (v) $\sum(x - 20)^2$ (vi) $\sum(x - 27)^2$

MEDIAN AND INTERQUARTILE RANGE

So far you have considered two measures of average: mode and mean, and two measures of variability: range and standard deviation.

When a set of data contains extreme values, known as outliers, the median is a more informative average and the interquartile range is a more useful measure of spread.

Median

The **median** is an average that is not influenced by extreme values.

For a set of n numbers arranged in ascending order:

when n is odd, the median is the middle value

when n is even, the median is the mean of the two middle values

This is summarised by saying that the **median** is the $\frac{1}{2}(n + 1)$th value.

Examples

(i) $n = 7$ 2   3   8 $\boxed{9}$ 10   14   23

              median = 9

When there are 7 numbers, the median is the 4th in the list.

(ii) $n = 6$ 145   152   160   164   173   185

$$\text{median} = \frac{160 + 164}{2} = 162$$

When there are 6 numbers, the median is the 3.5th value, i.e. the mean of the 3rd and 4th values.

Example 1.24

Find the median of each of these sets of numbers:

(i) 7, 7, 2, 3, 4, 2, 7, 9, 31

(ii) 36, 41, 27, 32, 29, 39, 39, 43

(i) *There are 9 values. Write them in ascending order.*

The median is the $\frac{1}{2}(9 + 1)$th value, i.e. the 5th value

     2   2   3   4 $\boxed{7}$ 7   7   9   31

             5th value

Notice that the median is not influenced by the extreme value of 31.

Median = 7.

(ii) *There are 8 values. Write them in ascending order.*

The median is the $\frac{1}{2}(8 + 1)$th value, i.e. the 4.5th value.

Find the mean of the 4th and 5th values.

    27   29   32 $\boxed{36 \quad 39}$ 39   41   43

             4.5th value

Median $= \frac{36 + 39}{2} = 37.5$

Quartiles

The lower and upper quartiles are not influenced by extreme values. They are values such that, together with the median, they split a distribution into four equal parts.

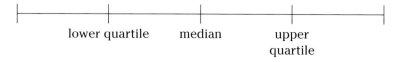

     lower quartile   median   upper quartile

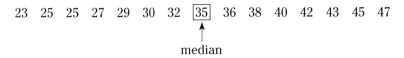

In this set of 15 numbers, the median is 35.

23 25 25 27 29 30 32 |35| 36 38 40 42 43 45 47

↑
median

To find the **lower quartile**, consider only the values **before** the median, and find the median of these.	To find the **upper quartile**, consider only the values **after** the median, and find the median of these.

23 25 25 |27| 29 30 32   36 38 40 |42| 43 45 47

↑   ↑
lower quartile   upper quartile

Showing the median and quartiles on one diagram, you now have

23 25 25 |27| 29 30 32 |35| 36 38 40 |42| 43 45 47

↑   ↑   ↑
lower quartile   median   upper quartile

Notation:
The lower quartile is often denoted by Q_1, the median by Q_2 and the upper quartile by Q_3.

For data arranged in order,

- the **lower quartile**, Q_1 is the median of all the values **before** the median
- the **upper quartile**, Q_3 is the median of all the values **after** the median.

Interquartile range

The difference between the quartiles gives the **interquartile range**. It tells you the range of the middle 50% of the data. It is a particularly useful measure of spread when there are outliers, as it is unaffected by one or two extreme values.

$$\text{Interquartile range} = \text{upper quartile} - \text{lower quartile} = Q_3 - Q_1$$

Don't forget to subtract.

As with the median, sometimes the quartiles are actual data values, sometimes they are not. Possible situations are illustrated in the following examples.

(i) $n = 11$

3 3 |5| 6 8 |9| 12 14 |19| 20 24

↑   ↑   ↑
Q_1 Q_2 Q_3
lower   median   upper
quartile   quartile

Lower quartile $Q_1 = 5$
Median $Q_2 = 9$
Upper quartile $Q_3 = 19$
Interquartile range $= 19 - 5 = 14$

(ii) $n = 6$

20 |23| 23 26 |27| 28

↑   ↑   ↑
Q_1 Q_2 Q_3

Lower quartile $Q_1 = 23$
Median $Q_2 = 24.5$
Upper quartile $Q_3 = 27$
Interquartile range $= 27 - 23 = 4$

(iii) $n = 8$

$$147 \quad 150 \underset{\underset{Q_1}{\uparrow}}{} 154 \quad 158 \underset{\underset{Q_2}{\uparrow}}{} 159 \quad 162 \underset{\underset{Q_3}{\uparrow}}{} 164 \quad 165$$

Lower quartile $Q_1 = 152$
Median $Q_2 = 158.5$
Upper quartile $Q_3 = 163$
Interquartile range $= 163 - 152 = 11$

(iv) $n = 9$

$$10 \quad 12 \underset{\underset{Q_1}{\uparrow}}{} 13 \quad 15 \quad \boxed{19} \underset{\underset{Q_2}{\uparrow}}{} 19 \quad 24 \underset{\underset{Q_3}{\uparrow}}{} 26 \quad 26$$

Lower quartile $Q_1 = 12.5$
Median $Q_2 = 19$
Upper quartile $Q_3 = 25$
Interquartile range $= 25 - 12.5 = 12.5$

Note: Sometimes the following rule is used to find the quartiles:

$$Q_1 = \tfrac{1}{4}(n + 1)^{\text{th}} \text{ value} \qquad\qquad Q_3 = \tfrac{3}{4}(n + 1)^{\text{th}} \text{ value}$$

This agrees with the method shown above when n is odd, but leads to discrepancies when n is even. To avoid problems, it is better to list the data and find the appropriate median, as in the examples above.

Example 1.25

To investigate hand-eye coordination in reacting to a stimulus, students took part in an experiment where a ruler was dropped and the distance it travelled before the student caught it was measured. The results of 21 girls and 27 boys are shown in the back-to-back stem-and-leaf diagram.

Distance, in cm

	Girls		Boys	
		4	8 9	(2)
		5	0 5 9	(3)
(1)	8	6	2 4	(2)
(5)	5 5 3 2 2	7	1 4 5	(3)
(2)	7 6	8	0 2 5 7 8 9	(6)
(6)	9 9 8 4 1 1	9	3 3 6	(3)
(3)	9 6 3	10	0 0 5 7 8	(5)
(3)	7 5 3	11	2 3	(2)
(1)	4	12	7	(1)

Key:
$8 \mid 6 \mid 2$ represents a distance of 6.8 cm for the girls and 6.2 cm for the boys.

Find the median and interquartile range for both sets of data. Comment on your answers.

<u>Boys</u>

There are 27 boys, so the median is the $\tfrac{1}{2}(27 + 1)^{\text{th}}$ value $= 14^{\text{th}}$ value.

Start counting from the top (4.8, 4.9, 5.0, 5.5, 5.9, 6.2, …) to the 14th value. Use the totals at the side as a guide.

Median $= 8.7$ cm

*To find the lower quartile, consider only the values **before** the median, and find the median of these.*

There are 13 values before the median, so the lower quartile Q_1 is the 7th value $= 6.4$ cm

Boys

4	8 9	(2)
5	0 5 9	(3)
6	2 $\boxed{4}$	(2)
7	1 4 5	(3)
8	0 2 5 $\boxed{7}$ 8 9	(6)
9	3 3 6	(3)
10	0 $\boxed{0}$ 5 7 8	(5)
11	2 3	(2)
12	7	(1)

*To find the upper quartile, consider only the values **after** the median, and find the median of these.*

There are 13 values after the median, so the upper quartile Q_3 is the 7th value after the median = 10.0 cm.

Interquartile range = $Q_3 - Q_1$ = 10.0 − 6.4 = 3.6 cm

Girls

There are 21 girls, so the median = $\frac{1}{2}(21 + 1)$th value = 11th value.

Start counting from the top (6.8, 7.2, 7.2, 7.3, …) to the 11th value.

Median = 9.4 cm · Take care when counting to the left of the stem.

Find the quartiles by finding the median of the data before and after the median.

There are 10 values before the median, so Q_1 is the 5.5th value (the mean of the 5th and 6th values). They are both the same, so Q_1 = 7.5 cm.

The 10 values before the median are
6.8, 7.2, 7.2, 7.3, 7.5, 7.5, 8.6, 8.7, 9.1, 9.1

Similarly, Q_3 is the 5.5th value after the median, i.e. the mean of the 5th and 6th values, so $Q_3 = \frac{1}{2}(10.6 + 10.9)$ = 10.75 cm

The 10 values after the median are
9.8, 9.9, 9.9, 10.3, 10.6, 10.9, 11.3, 11.5, 11.7, 12.4

Interquartile range = 10.75 − 7.5 = 3.25 cm

Girls

		Stem
	4	
	5	
(1)	8	6
(5)	5 5 3 2 2	7
(2)	7 6	8
(6)	9 9 8 4 1 1	9
(3)	9 6 3	10
(3)	7 5 3	11
(1)	4	12

Summary

	Distance (cm)	
	Girls	**Boys**
Median	9.4	8.7
Interquartile range	3.25	3.6

The distance the ruler travelled was further on average for the girls, indicating that the girls were slower to react.

However, the larger interquartile range for the boys indicates that their results were more variable. The girls' results were more consistent.

CUMULATIVE FREQUENCY

Cumulative frequency is the total frequency up to a particular item. Cumulative frequency is particularly useful when finding the median and quartiles.

Cumulative frequency − ungrouped data

A survey was carried out to find the number of attempts needed by candidates to pass the driving test. The results for 99 candidates at a particular test centre are shown in the frequency table.

Number of attempts	1	2	3	4	5	6	
Frequency	32	42	13	6	4	2	Total 99

To find the median and quartiles you could list all the observations in order:

1, 1, 1, ..., 1, 2, 2, ..., 2, 3, ..., 6, 6.

However, this would be very tedious. It is easier to form a cumulative frequency table and use it to find the median and quartiles.

Cumulative frequency table

Number of attempts	$\leqslant 1$	$\leqslant 2$	$\leqslant 3$	$\leqslant 4$	$\leqslant 5$	$\leqslant 6$
Cumulative frequency	32	74	87	93	97	**99**

32 + 42   32 + 42 + 13   This is the total
           or 74 + 13   number of candidates.

Median

There are 99 observations, so the median is the $\frac{1}{2}(99 + 1)^{\text{th}}$ observation, i.e. the 50th observation.

32 candidates needed 1 attempt and 74 candidates needed 2 or fewer attempts, so the median number of attempts is 2.

Quartiles

As there are 49 values before the median, the lower quartile is the $\frac{1}{2}(49 + 1)^{\text{th}} = 25^{\text{th}}$ observation, so the lower quartile Q_1 is 1.

As there are 49 values after the median, the upper quartile is the 25th observation after the median, i.e. the 75th observation, so the upper quartile Q_3 is 3.

The 75th observation is the first '3' to appear in the list when all the observations are written in order.

Interquartile range $= Q_3 - Q_1$
$$= 3 - 1$$
$$= 2 \text{ attempts}$$

Exercise 1g

1 These are the test marks of 11 students.

52, 61, 78, 49, 47, 79, 54, 58, 62, 73, 72

Find

(i) the median,

(ii) the lower quartile,

(iii) the upper quartile,

(iv) the interquartile range.

2 For each of the following sets of numbers, find

(i) the median,

(ii) the interquartile range.

(a) 4, 6, 18, 25, 9, 16, 22, 5, 20, 4, 8, 15, 9, 13, 10

(b) 192, 217, 189, 210, 214, 204

(c) 1267, 1896, 895, 3457, 2164, 2347, 2347, 2045

(d) 0.7, 0.4, 0.65, 0.78, 0.45, 0.32, 1.9, 0.0078, 0.54, 1.32

(e) 0.3, -1.5, -3.5, -3.05, 1.4, -2.6, -0.02

3 The table shows the scores obtained when a die is thrown 60 times.

Score	1	2	3	4	5	6
Frequency	12	9	8	13	9	9

Find

   (i) the median score,

   (ii) the lower quartile and upper quartile,

   (iii) the interquartile range.

4 Find the median and interquartile range of the distributions represented by the stem-and-leaf diagrams:

   (i)
Stem	Leaf	
1	0 5	(2)
2	3 4 4	(3)
3	2 8 8	(3)
4	1 5 6 6 7	(5)
5	2 3 3	(3)
6	5 7 8 8	(4)
7	2 4	(2)
8	0	(1)

Key: 5 | 2 means 5.2

   (ii)
Stem	Leaf
12	3 4 3 9
13	2 2 3 4 7 8 8 9 9
14	0 3 4 4 7
15	1 2

Key: 12 | 3 means 0.123

5
x	5	6	7	8	9	10
f	6	11	15	18	6	5

For the above frequency distribution, find

   (i) the mode,

   (ii) the median,

   (iii) the mean.

6
x	12	13	14	15	16
f	3	9	11	15	17

For the above frequency distribution, find

   (i) the range,

   (ii) the interquartile range,

   (iii) the standard deviation.

7 The vertical line graph shows the number of goals scored in a match by Jemima in the 24 games last season.

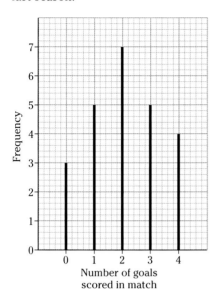

(i) Find the median number of goals.

(ii) Find the interquartile range.

8 In a survey on the number of absences in the term of the 32 children in a class, the data were recorded in a cumulative frequency table.

Times absent	0	$\leqslant 1$	$\leqslant 2$	$\leqslant 3$	$\leqslant 4$	$\leqslant 5$	$\leqslant 6$	$\leqslant 7$
Cumulative frequency	5	11	20	23	27	28	31	32

(i) Find the median number of absences.

(ii) Find the interquartile range.

(iii) Copy and complete this frequency table.

Times absent	0	1	2	3	4	5	6	7
Frequency								

(iv) Calculate the mean number of absences per child.

(v) Calculate the standard deviation.

9 The stem-and-leaf diagram below represents data collected for the number of hits on an internet site on each day in March 2007. There is one missing value, denoted by x.

```
0 │ 0 1 5 6              (4)
1 │ 1 3 5 6 6 8          (6)
2 │ 1 1 2 3 4 4 4 8 9    (9)
3 │ 1 2 2 2 x 8 9        (7)
4 │ 2 5 6 7 9            (5)
```

Key: 1 | 5 represents 15 hits

(i) Find the median and lower quartile for the number of hits each day.

(ii) The interquartile range is 19. Find the value of x.   Cambridge Paper 6 Q1 J08

Cumulative frequency – grouped data

When data have been grouped, cumulative frequencies are calculated up to each **upper class boundary**, as in the following example.

The <u>frequency</u> table shows the heights of broad bean plants six weeks after planting.

Height, x cm	$3 \leqslant x < 6$	$6 \leqslant x < 9$	$9 \leqslant x < 12$	$12 \leqslant x < 15$	$15 \leqslant x < 18$	$18 \leqslant x < 21$	
Frequency	1	2	11	10	5	1	Total **30**

The upper class boundaries are 6, 9, 12, 15, 18 and 21.

The lower boundary of the first interval is 3. It is a good idea to include this in the cumulative frequency table.

This is the <u>cumulative frequency</u> table

Height, x cm	< 3	< 6	< 9	< 12	< 15	< 18	< 21
Cumulative frequency	0	1	3	14	24	29	**30**

$\uparrow$ $1 + 2$ $\uparrow$ $1 + 2 + 11$ $\uparrow$ This must be the same as the total frequency.

Cumulative frequencies can be thought of as 'running totals' of the frequencies.

The cumulative frequency table can be illustrated on a **cumulative frequency graph** in which the cumulative frequencies are plotted against the **upper class boundaries**. The points are joined either with a curve or with straight lines. Often curves are drawn, but either method is acceptable in the examination.

Cumulative frequency curve to show heights of 30 broad bean plants

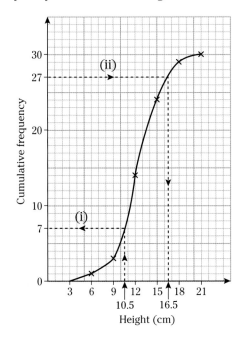

Values can be **estimated** from the graph, for example:

(i) Estimate how many plants had a height less than 10.5 cm.
- From 10.5 on the *x*-axis (height) draw a vertical line up to the curve.
- Now draw a horizontal line to the *y*-axis (cumulative frequency) and read off the value.

From the graph, 7 plants had a height less than 10.5 cm.

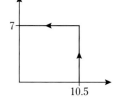

(ii) 10% of plants had a height of at least *x* cm. Estimate *x*.
- 10% of 30 = 3
  3 plants had a height of at least *x* cm, so 27 plants had a height less than *x* cm.
- From 27 on the *y*-axis draw a horizontal line to the curve.
- Now draw a vertical line down to the *x*-axis and read off the value.

From the graph, 10% of the plants had a height of at least 16.5 cm.

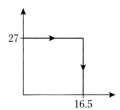

Example 1.26

A factory produces certain components. In a quality control test, 500 components were weighed and their weights recorded to the nearest gram. The table shows the results.

Weight (g)	60–69	70–74	75–79	80–84	85–89
Frequency	30	90	130	210	40

(i) Construct a cumulative frequency table and draw a cumulative frequency graph.
(ii) Components that weigh less than 64.5 grams or more than 87.5 grams were rejected. Use your graph to estimate the percentage of components that were accepted.

(i) The weights have been recorded to the nearest gram, so the upper class boundaries are 69.5, 74.5, 79.5, 84.5, 89.5.

The lower boundary of the first interval is 59.5.

Show the boundaries and the cumulative frequencies in a table before drawing the curve.

Cumulative frequency table

Weight (g)	< 59.5	< 69.5	< 74.5	< 79.5	< 84.5	< 89.5
Cumulative frequency	0	30	120	250	460	500

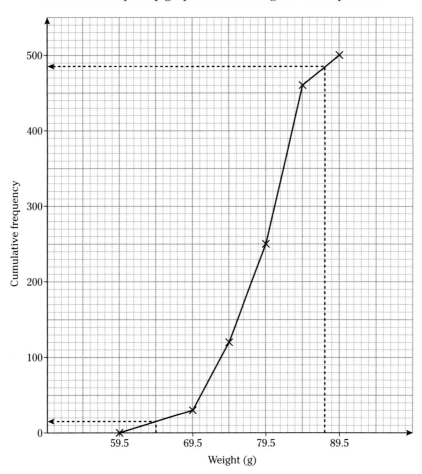

Cumulative frequency graph to show weights of components

(ii) *Find 64.5 on the x-axis. Draw a vertical line up to the curve, then a horizontal line to the y-axis. Read off the value. Repeat with 87.5 on the x-axis.*

When you are asked to estimate values from the graph, you must show your method by drawing lines on the graph.

Number of components weighing less than 64.5 g = 15

Number of components weighing less than 87.5 g = 485

Number of satisfactory components = 485 − 15 = 470

Percentage of satisfactory components = $\frac{470}{500} \times 100 = 94\%$

Example 1.27

The cumulative frequency table shows the times taken by students to travel to college on a particular day.

Time (minutes)	< 10	< 15	< 20	< 30	< 45
Cumulative frequency	35	79	157	350	400

Construct a frequency table using intervals $0 \leqslant t < 10$, $10 \leqslant t < 15$, $15 \leqslant t < 20$, $20 \leqslant t < 30$, $30 \leqslant t < 45$ and use it to estimate the mean time taken to travel to college on that day.

Calculate the frequencies in each interval as follows.

$0 \leqslant t < 10$	frequency $= 35$
$10 \leqslant t < 15$	frequency $= 79 - 35 = 44$
$15 \leqslant t < 20$	frequency $= 157 - 79 = 78$
$20 \leqslant t < 30$	frequency $= 350 - 157 = 193$
$30 \leqslant t < 45$	frequency $= 400 - 350 = 50.$

Check that this gives a total of 400.

Show the frequencies in a frequency table. To estimate the mean you will also need the mid-interval values.

Time (minutes)	Mid-interval value, t	f	$t \times f$
$0 \leqslant t < 10$	5	35	175
$10 \leqslant t < 15$	12.5	44	550
$15 \leqslant t < 20$	17.5	78	1365
$20 \leqslant t < 30$	25	193	4825
$30 \leqslant t < 45$	37.5	50	1875
		$\sum f = 400$	$\sum (t \times f) = 8790$

$$\bar{t} = \frac{\sum (t \times f)}{\sum f} = \frac{8790}{400} = 21.975 = 22.0 \text{ mins (3 s.f.)}$$

If you use the statistical functions on your calculator, it is a good idea to state the mid-interval values you are using.

Finding the median and quartiles of grouped data

When data have been grouped into intervals, the original information has been lost, so it is only possible to make **estimates** of the median and quartiles. One way of doing this is to read off the appropriate values from a cumulative frequency graph.

The following cumulative frequency curve shows the times spent by a group of 23 students on their statistics assignment.

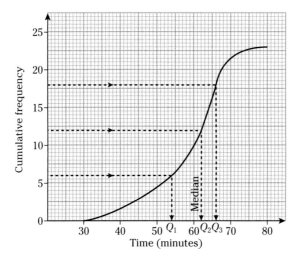

There are 23 students.

Median $Q_2 = \frac{1}{2}(23 + 1)^{\text{th}}$ value $= 12^{\text{th}}$ value.

Q_1 is the median of the 11 values before the median $= 6^{\text{th}}$ value.

Q_3 is the median of the 11 values after the median $= (12 + 6)^{\text{th}}$ value $= 18^{\text{th}}$ value.

Using the graph, estimates are as follows:

      median $Q_2 = 62$ minutes,

      lower quartile $Q_1 = 54$ minutes,

      upper quartile $Q_3 = 66$ minutes

Large data sets

Often cumulative frequency graphs are drawn to represent large sets of data. If, for example, you have 100 observations, the median is the 50.5^{th} value, but the scale is probably such that there is very little difference between reading off the 50.5^{th} value and the 50th value. So the following rule is usually used when n is large.

For grouped data, when n is large:

$$\text{median} = \tfrac{1}{2}n^{\text{th}} \text{ value}$$
$$\text{lower quartile} = \tfrac{1}{4}n^{\text{th}} \text{ value}$$
$$\text{upper quartile} = \tfrac{3}{4}n^{\text{th}} \text{ value}$$

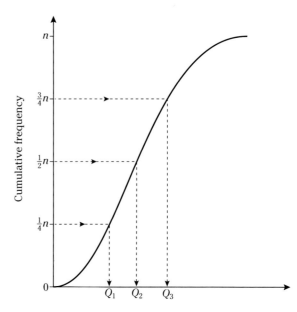

Example 1.28

The cumulative frequency table gives the heights of 400 children in a certain school.

Height, x cm	< 100	< 110	< 120	< 130	< 140	< 150	< 160	< 170
Cumulative frequency	0	27	85	215	320	370	395	400

(i) Draw a cumulative frequency curve.

(ii) Use the curve to estimate the median height.

(iii) Determine the interquartile range.

(i) Cumulative frequency curve to show heights of schoolchildren.

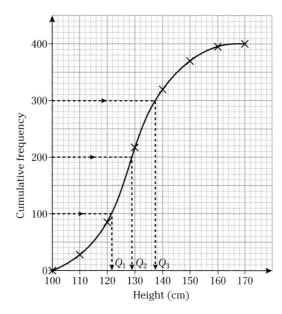

Since n is large:

(ii) Median $Q_2 = \frac{1}{2}(400)^{\text{th}}$ value

$\qquad = 200^{\text{th}}$ value

$\qquad = 129 \, \text{cm}$

(iii) $Q_1 = \frac{1}{4}(400)^{\text{th}}$ value

$\qquad = 100^{\text{th}}$ value

$\qquad = 121.5 \, \text{cm}$

$Q_3 = \frac{3}{4}(400)^{\text{th}}$ value

$\qquad = 300^{\text{th}}$ value

$\qquad = 137.5 \, \text{cm}$

Interquartile range (IQR)

$\qquad = Q_3 - Q_1$

$\qquad = 137.5 - 121.5$

$\qquad = 16 \, \text{cm}$

Example 1.29

The arrival times of 204 trains were noted and the number of minutes, t, that each train was late was recorded.

Number of minutes late (t)	$-2 \leqslant t < 0$	$0 \leqslant t < 2$	$2 \leqslant t < 4$	$4 \leqslant t < 6$	$6 \leqslant t < 10$
Number of trains	43	51	69	22	19

(i) Explain what $-2 \leqslant t < 0$ means about the arrival times of the trains.

(ii) Draw a cumulative frequency graph, and from it estimate the median and interquartile range of the number of minutes late of these trains.

Cambridge Paper 6 Q5 N07

(i) Some trains arrived up to 2 minutes **early**.

(ii) Cumulative frequency table

You are advised to show the cumulative frequencies in a table before you plot them. If you make a small slip, you could still get method marks in the examination.

Number of minutes late (t)	$t < -2$	$t < 0$	$t < 2$	$t < 4$	$t < 6$	$t < 10$
Cumulative frequency	0	43	94	163	185	**204**

Make sure that this gives the total number of trains.

Cumulative frequency curve to show lateness of trains

Since n is large,

$$\text{median} = \tfrac{1}{2}(204)^{\text{th}} \text{ value}$$
$$= 102^{\text{nd}} \text{ value}$$
$$= 2.2 \text{ minutes}$$

$$Q_1 = \tfrac{1}{4}(204)^{\text{th}} \text{ value}$$
$$= 51^{\text{st}} \text{ value}$$
$$= 0.3 \text{ minutes}$$

$$Q_3 = \tfrac{3}{4}(204)^{\text{th}} \text{ value}$$
$$= 153^{\text{rd}} \text{ value}$$
$$= 3.6 \text{ minutes}$$

Interquartile range
$$= Q_3 - Q_1$$
$$= 3.6 - 0.3$$
$$= 3.3 \text{ minutes}$$

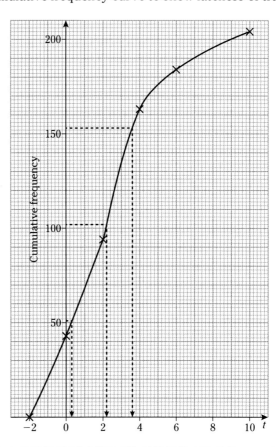

Example 1.30

The times, to the nearest minute, taken by 120 students to write a timed essay were recorded. The results are shown in the table.

Time (minutes)	40–44	45–49	50–54	55–59	60–64
Frequency	8	24	32	30	26

(i) Construct the cumulative frequency table and draw a cumulative frequency graph.

(ii) Use your graph to estimate the lower quartile and the median.

Another group of 40 students wrote the same essay and all of them took at least 1 hour to complete it.

(iii) Use your graph to estimate the lower quartile of all 160 students.

(iv) Explain why it is not possible to estimate the interquartile range of the times spent by all 160 students.

(i) The upper class boundaries are 44.5, 49.5, 54.5, 59.5, 64.5
The lower class boundary of the first interval is 39.5

Cumulative frequency table:

Time (t minutes)	$t < 39.5$	$t < 44.5$	$t < 49.5$	$t < 54.5$	$t < 59.5$	$t < 64.5$
Cumulative frequency	0	8	32	64	94	120

Cumulative frequency curve to show times

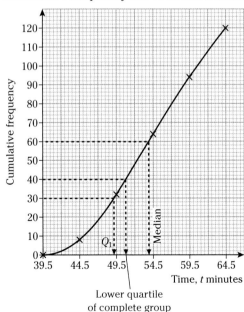

Lower quartile
of complete group

(ii) Since n is large,

$$\text{Median} = \tfrac{1}{2}(120)^{\text{th}} \text{ value}$$
$$= 60^{\text{th}} \text{ value}$$
$$= 54 \text{ minutes}$$
$$Q_1 = \tfrac{1}{4}(120)^{\text{th}} \text{ value}$$
$$= 30^{\text{th}} \text{ value}$$
$$= 49 \text{ minutes}$$

Since all in the second group took at least 60 minutes, the cumulative frequency curve for 160 students is the same as the curve for 120 students as far as (59.5, 94).

(iii) Lower quartile of 160 students
$$= \tfrac{1}{4}(160)^{\text{th}} \text{ value}$$
$$= 40^{\text{th}} \text{ value}$$
$$= 50.5 \text{ minutes}$$

> Note that you are able to find the lower quartile of 160 students without knowing the full set of data.

(iv) To find the interquartile range, the upper quartile of the **complete set of times** is needed $= \tfrac{3}{4}(160)^{\text{th}}$ value $= 120^{\text{th}}$ value.

However, after a time of 59.5 (the 94th value), the curves are not the same, so it is not possible to find the 120th value of the complete set of times without further information.

Exercise 1h

1 The cumulative frequency curve has been drawn from information about the amount of time spent by 50 people in a supermarket on a particular day.

Use the graph to estimate

(i) how many people spent at least 17 minutes but less than 27 minutes in the supermarket,

(ii) the value of t, where 60% of the people spent less than t minutes in the supermarket,

(iii) the value of s, where 60% of the people spent at least s minutes in the supermarket,

(iv) the median time,

(v) the interquartile range.

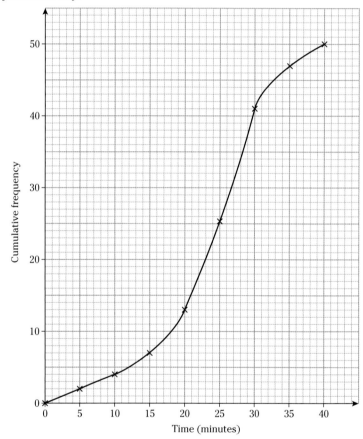

2 Eggs laid at Hill Farm are weighed and the results grouped as shown:

Mass (x grams)	$46 \leqslant x < 50$	$50 \leqslant x < 54$	$54 \leqslant x < 58$	$58 \leqslant x < 62$	$62 \leqslant x < 66$	$66 \leqslant x < 70$	$70 \leqslant x < 74$
Frequency	6	4	10	24	20	12	4

(i) Construct a cumulative frequency table and draw a cumulative frequency graph.

(ii) Estimate the median mass.

(iii) The mass of three-quarters of the eggs was at least m grams. Estimate m from the graph.

3 Fifty soil samples were collected in an area of woodland, and the pH value for each sample was found. The cumulative frequency distribution was constructed as shown in the table.

pH value	< 4.4	< 4.8	< 5.2	< 5.6	< 6.0	< 6.4	< 6.8	< 7.2	< 7.6	< 8.0	< 8.4
Cumulative frequency	0	1	2	5	10	19	38	43	46	49	50

(i) Draw a cumulative frequency curve.

(ii) Estimate the percentage of samples with a pH value less than 7.

(iii) Estimate the median.

(iv) Taking equal width intervals of $4.4 \leqslant x < 4.8$, $4.8 \leqslant x < 5.2$, etc., construct the frequency distribution and draw a histogram. Show the median on the histogram.

4 In a quality-control survey the length of life, measured to the nearest hour, of 100 light bulbs is noted. The results are summarised in the table.

Length of life (to nearest hour)	650–669	670–679	680–689	690–699	700–729
Frequency	6	14	40	34	6

(i) Draw a cumulative frequency graph and use it to estimate the median and interquartile range.

(ii) Estimate the mean and standard deviation.

5 The weekly maximum temperatures in a certain town were recorded, to the nearest °C, over a period of two years and grouped in the following table.

Temperature (x °C)	−5 to −1	0–4	5–9	10–14	15–19	20–24	25–29
Frequency	8	12	17	31	23	9	4

(i) State the boundaries and width of the interval 0–4.

(ii) Draw the cumulative frequency graph.

(iii) Use your graph to estimate the median temperature.

(iv) A week is classified as 'extremely warm' when the weekly maximum is at least 21 °C. Use your graph to estimate the percentage of weeks that are classified as 'extremely warm'.

6 The distribution of the times taken when a certain task was performed by a large group of people was noted. It was found that 20% performed the task in less than 30 minutes, 40% in less than 38 minutes, 60% in less than 45 minutes and 80% in less than 53 minutes. The shortest time was 10 minutes and the greatest time was 69 minutes.

(i) Draw a cumulative frequency graph to illustrate the data.

(ii) Estimate the median time.

(iii) Estimate the percentage of people who performed the task in less than 50 minutes.

7   A survey was made of the times taken by a group of people to complete an assault course. The results are shown in the histogram.

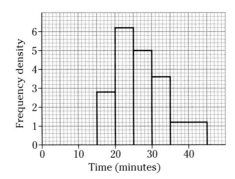

(i)   Copy and complete the frequency table.

Time (x minutes)	$15 \leqslant x < 20$	$20 \leqslant x < 25$	$25 \leqslant x < 30$	$30 \leqslant x < 35$	$35 \leqslant x < 45$
Frequency		31			

(ii)   Which interval does the median lie in?

(iii)   Draw a cumulative frequency graph.

(iv)   Use your graph to estimate the median and interquartile range.

8   Each of 200 sportsmen was asked to state the distance, x km, he needs to travel to obtain access to suitable training facilities. The results are summarised in the table below.

Distance (x km)	$0 \leqslant x < 4$	$4 \leqslant x < 10$	$10 \leqslant x < 20$	$20 \leqslant x < 35$	$35 \leqslant x < 60$
Frequency	5	10	39	95	51

(a)   (i)   Construct a cumulative frequency table and draw a cumulative frequency graph.

Use your graph to estimate

(ii)   the median distance,

(iii)   the interquartile range of the distances,

(iv)   the percentage of sportsmen who need to travel 30 km or more.

(b)   (i)   Draw a histogram to represent the data.

(ii)   Estimate the mean and standard deviation of the distances travelled.

9   In a survey, two groups of 200 students were asked to keep a record of the number of text messages they sent during a certain period of time.

The results are as follows.

Number of text messages sent	1–5	6–10	11–15	16–25	26–34	35–40
Group 1: Frequency	4	8	11	35	104	38
Group 2: Frequency	11	17	27	87	50	8

(i)   On the same diagram, draw two cumulative frequency graphs to represent the data.

(ii)   For each group, estimate the median and the interquartile range.

(iii)   Compare the two groups.

10 In a recent survey, 640 people were asked about the length of time each week that they spent watching television. The median time was found to be 20 hours, and the lower and upper quartiles were 15 hours and 35 hours respectively. The least amount of time that anyone spent was 3 hours, and the greatest amount was 60 hours.

   (i) On graph paper, show these results using a fully labelled cumulative frequency graph.

   (ii) Use your graph to estimate how many people watched more than 50 hours of television each week.

Cambridge Paper 6 Q2 J04

11 The manager of a company noted the times spent in 80 meetings. The results were as follows.

Time (t minutes)	$0 < t \leqslant 15$	$15 < t \leqslant 30$	$30 < t \leqslant 60$	$60 < t \leqslant 90$	$90 < t \leqslant 120$
Frequency	4	7	24	38	7

Draw a cumulative frequency graph and use this to estimate the median time and the interquartile range.

Cambridge Paper 6 Q2 J02

BOX-AND-WHISKER PLOTS

In a **box-and-whisker plot** the median and quartiles are shown, as well as the minimum and maximum values of a distribution. It gives a very good visual summary of a distribution and is particularly useful when comparing sets of data.

A survey on the heights of all the girls in a particular year group in a school gave the following information.

Minimum height	144 cm
Lower quartile	159 cm
Median	165 cm
Upper quartile	169 cm
Maximum height	181 cm

This information is shown on the box-and-whisker plot below. The ends of the whiskers are placed at the minimum and maximum values, the box is drawn at the lower and upper quartiles and the median is drawn in the box.

The width of the box gives the interquartile range and the width from the minimum value to the maximum value gives the range.

Box-and-whisker plot to show heights of 15-year-old girls

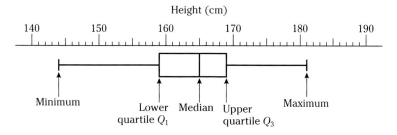

Tips on drawing the box-and-whisker plot

The scale must be linear and labelled.

The whiskers must not be drawn through the box.

The height of the box is a matter for personal choice, although it is more informative if it is not too tall.

When comparing two or more box-and-whisker plots, choose the same height for the box in each plot.

A box-and-whisker plot illustrates the spread of a distribution and also gives an idea of the shape of the distribution.

Negative skew

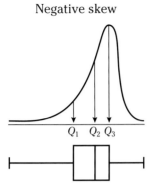

The whisker to the left is longer and the median is nearer to Q_3.

Symmetrical

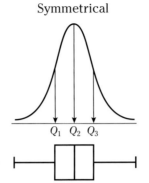

The whiskers are of equal length and the median is in the middle of the box.

Positive skew

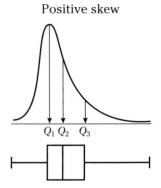

The whisker to the right is longer and the median is nearer to Q_1.

Note that the ideas of skew and skewness are not required in the examination.

Example 1.31

Two groups of people played a computer game which tested how quickly they reacted to a visual instruction to press a particular key. The computer measured their reaction times in seconds, to the nearest tenth of a second. The following summary statistics were displayed for each group.

	Minimum	Lower quartile Q_1	Median Q_2	Upper quartile Q_3	Maximum
Group 1	0.6	0.8	1.0	1.5	1.9
Group 2	0.4	0.7	1.0	1.3	1.6

Draw two box-and-whisker plots and compare the reaction times of the two groups.

Box-and-whisker plots to show reaction times

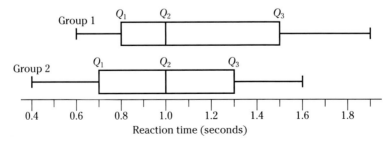

The median reaction time is 1.0 seconds for both groups. However, the range of times for Group 2 is smaller than for Group 1 and the times are evenly distributed for Group 2.

There is a greater spread of times for Group 1. Their distribution has a long tail to the right, indicating that there are a few very high values.

In general, Group 2 has the faster reaction time.

Example 1.32

The following back-to-back stem-and-leaf diagram shows the cholesterol count for a group of 45 people who exercise daily and for another group of 63 who do not exercise. The figures in brackets show the number of people corresponding to each set of leaves.

People who exercise		People who do not exercise	
(9)	9 8 7 6 4 3 2 2 1 | 3	1 5 7 7	(4)
(12)	9 8 8 8 7 6 6 5 3 3 2 2 | 4	2 3 4 4 5 8	(6)
(9)	8 7 7 7 6 5 3 3 1 | 5	1 2 2 2 3 4 4 5 6 7 8 8 9	(13)
(7)	6 6 6 6 4 3 2 | 6	1 2 3 3 3 4 5 5 5 7 7 8 9 9	(14)
(3)	8 4 1 | 7	2 4 5 5 6 6 7 8 8	(9)
(4)	9 5 5 2 | 8	1 3 3 4 6 7 9 9 9	(9)
(1)	4 | 9	1 4 5 5 8	(5)
(0)	| 10	3 3 6	(3)

> **Key:** 2 | 8 | 1 represents a cholesterol count of 8.2 in the group who exercise and 8.1 in the group who do not exercise.

(i) Give one useful feature of a stem-and-leaf diagram.

(ii) Find the median and quartiles of the cholesterol count for the group who do not exercise.

You are given that the lower quartile, median and upper quartile of the cholesterol count for the group who exercise are 4.25, 5.3 and 6.6 respectively.

(iii) On a single diagram on graph paper, draw two box-and-whisker plots to illustrate the data.

Cambridge Paper 6 Q4 J05

(i) A stem-and-leaf diagram shows all the data. *Other advantages include being able to see the shape of the distribution and being able to find the mode easily.*

(ii) **People who do not exercise**
There are 63 people who do not exercise, so the median is the $\frac{1}{2}(63 + 1)^{\text{th}}$ value = 32nd value.

People who do not exercise

3	1 5 7 7	(4)
4	2 3 4 4 5 8	(6)
5	1 2 2 2 3 [4] 4 5 6 7 8 8 9	(13)
6	1 2 3 3 3 4 5 5 [5] 7 7 8 9 9	(14)
7	2 4 5 5 6 6 7 8 8	(9)
8	1 [3] 3 4 6 7 9 9 9	(9)
9	1 4 5 5 8	(5)
10	3 3 6	(3)

Start counting from the top (3.1, 3.5, 3.7, 3.7, 4.2, ...) until you reach the 32nd value which is 6.5. Note that, using the numbers in brackets at the side as a guide, you can see that you will be at the 23rd value when you get to 5.9, so you could just count on from there (6.1, 6.2, ...) to get the 32nd value.

Median = 6.5

Find the quartiles by finding the median of the values either side of the median.

There are 31 values before the median, so the lower quartile is the 16th value.

Lower quartile = 5.4

There are 31 values after the median, so the upper quartile is the 16th value after the median.

Count on after the median until you get to the 16th value.

Alternatively, you could count back from the highest value until you get to the 16th, i.e. 10.6, 10.3, 10.3, 9,8, …, 8.3

Upper quartile = 8.3

(iii) *Note the ends of the whiskers.*

People who exercise: minimum 3.1, maximum 9.4
People who do not exercise: minimum 3.1, maximum 10.6

Box-and-whisker plots to show cholesterol counts

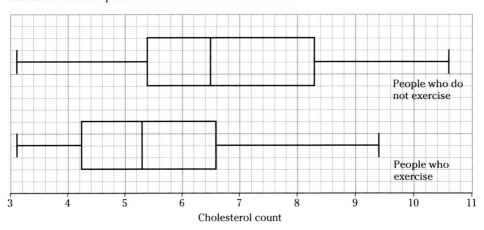

Hints on drawing the box-and-whisker diagrams:

Use graph paper, as instructed.

Make sure that your horizontal scale is linear. It needs to include 3.1 and 10.6, so a scale from 3 to 11 would be suitable.

Remember to label the horizontal scale 'cholesterol count'.

Label each box plot so that it can be seen at a glance which is which.

Example 1.33

In a survey, people were asked how long they took to travel to and from work, on average. The median time was 3 hours 36 minutes, the upper quartile was 4 hours 42 minutes and the interquartile range was 3 hours 48 minutes. The longest time was 5 hours 12 minutes and the shortest time was 30 minutes.

(i) Find the lower quartile.

(ii) Represent the information by a box-and-whisker plot, using a scale of 2 cm to represent 60 minutes.

Cambridge Paper 6 Q3 N06

(i) Interquartile range = Upper quartile − lower quartile
    So, 3 hr 48 min = 4 hr 42 min − lower quartile
    lower quartile = 4 hr 42 min − 3 hr 48 min
                   = 54 min

You could work in hours, with the minutes converted to decimal parts of an hour, where 6 minutes = 0.1 hours.

(ii) Box-and-whisker plot to show travelling times

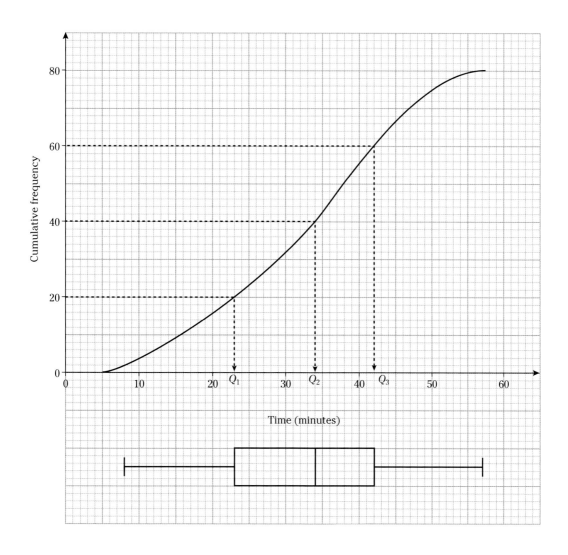

Remember to label the axis.

Take care with the scale. If your graph paper has 10 small squares to 2 cm, then 1 small square represents 0.1 hours = 6 minutes.

A box-and-whisker plot can be drawn on the same diagram as a cumulative frequency graph.

In the illustration below, the cumulative frequency graph shows the times taken to travel to work by 80 employees in a firm. The median (Q_2) and quartiles (Q_1 and Q_3) are shown on the graph. With the additional information that the shortest time is 8 minutes and the greatest time is 57 minutes, the box-and-whisker plot is drawn immediately underneath, using the same horizontal scale.

This technique can be particularly useful when comparing two distributions. The cumulative frequency curves below show the marks of two groups of 120 people in a test. A common mistake when looking at these curves is to say that Group 1 did better than Group 2 because the curve for Group 1 is 'higher than' or 'above' the curve for Group 2. However, when you look at the box-and-whisker plots for each group, you can see that this is not the case. The marks for Group 2 were much higher.

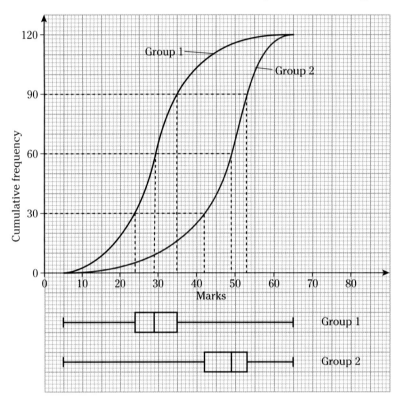

Outliers

Sometimes unusually high or low values occur in a set of data. These extreme values are called **outliers**. One of the ways in which outliers are identified uses the interquartile range, as follows:

'An outlier can be defined as a value that is more than $1.5 \times$ IQR below the lower quartile or above the upper quartile.'

> If you are asked to identify outliers in the examination, you will be given the rule to use.

The interquartile range is given by the width of the box in a box-and-whisker plot. You can have a good idea of whether there are outliers just by looking at the length of the whiskers in relation to the width of the box.

Consider this set of numbers, illustrated by the box-and-whisker plot below:

$$2 \quad 14 \quad 14 \quad \boxed{18} \quad 18 \quad 19 \quad 19 \quad \boxed{20} \quad 21 \quad 22 \quad 22 \quad \boxed{26} \quad 29 \quad 39 \quad 40$$
$$ Q_1 Q_2 Q_3$$

Minimum = 2
Maximum = 40
Median = 20
Lower quartile $Q_1 = 18$
Upper quartile $Q_3 = 26$

Width of box = IQR = $Q_3 - Q_1 = 26 - 18 = 8$

Lower tail: $Q_1 - 1.5 \times 8 = 18 - 12 = 6$, so any value below 6 is an outlier

Upper tail: $Q_3 + 1.5 \times 8 = 26 + 12 = 38$, so any value above 38 is an outlier

Hence the outliers are **2**, **39** and **40**.

A 'refined box plot' can be drawn as follows. The outliers are shown with a cross.

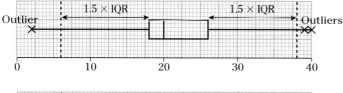

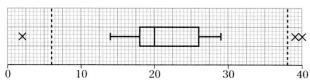

Note: You will not be asked to draw refined box plots in the examination. They are included here for interest.

Exercise 1i

1 Below is a box-and-whisker plot of pupils' marks in a test.

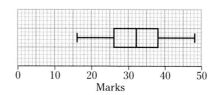

   (i) Write down the lower and upper quartiles.

  (ii) Write down the median.

 (iii) Work out the range.

 (iv) Work out the interquartile range.

  (v) Comment on the shape of the distribution.

2 In a survey on the number of words in a sentence in a particular novel, 100 sentences were chosen at random and the number of words in the sentence noted. The results are summarised in the table.

Lowest value	Lower quartile	Median	Upper quartile	Highest value
4	15	20	32	45

Draw a box-and-whisker plot on graph paper.

3 In each of the following, draw a box-and-whisker plot on graph paper to illustrate the following data and comment on the shape of the distribution.

   (i) 3, 5, 10, 11, 12, 16, 17, 17, 19, 20, 22

  (ii) 96, 105, 123, 151, 167, 178, 185, 200, 202, 220, 238, 246, 252, 269, 297

 (iii) −7, −5, −4, −4, −3, −3, −2, −1, 0, 1, 4, 6, 8, 9

4 The following stem-and-leaf diagram summarises the blood glucose level of a patient measured daily over a period of time.

Blood glucose level

| Key: 5 | 2 means 5.2 |

```
5 | 2 2 3 4 7 7 9 9    (8)
6 | 0 3 3 6 9          (5)
7 | 1 4 7 8 8          (5)
8 | 2 3 7              (3)
9 | 0 4                (2)
```

(i) Find the median and quartiles of these data.

(ii) On graph paper, draw a box-and-whisker plot to represent the data.

5 These are the times of the postal delivery to my house over four successive weeks.

```
9:01   9:22   9:30   9:19   9:15   9:29
9:45   9:53   9:02   9:05   9:31   9:47
9:17   9:48   9:29   9:09   9:29   9:02
9:10   9:12   9:25   9:10   9:13   9:19
```

(i) Draw a stem-and-leaf diagram.

(ii) Find the median time.

(iii) Find the quartiles.

(iv) Draw a box-and-whisker plot.

6 Draw a box-and-whisker plot to represent the distribution given in the table.

x	0	1	2	3	4
f	4	6	8	6	1

7

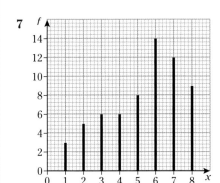

For the data represented in the vertical line graph above:

(i) State the minimum and maximum values of x.

(ii) Construct a cumulative frequency distribution table.

(iii) Find the median and the quartiles.

(iv) Draw a box-and-whisker plot.

8   In an experiment, 21 girls estimated the length of a line and gave their answers in millimetres. Their results were as shown.

       51   45   31   43   97   16   18   23   34   35   35
       85   62   20   22   51   57   49   22   18   27

   (i) Find the median, quartiles and interquartile range (IQR).

   (ii) Draw a box-and-whisker plot.

An outlier can be defined as a value that is more than $1.5 \times$ IQR below the lower quartile or above the upper quartile.

   (iii) Identify any outliers using this rule.

9   This back-to-back stem-and-leaf diagram gives the daily hours of sunshine in December and July in a certain city.

   Key: $1 \mid 3 \mid 5$ means 3.1 hours in December and 3.5 hours in July

                              Hours of sunshine
                  December   **July**
   (12)   8 8 7 3 3 2 0 0 0 0 0 0 │ 0 │ 3 4   (2)
   (6)   9 8 8 6 0 0 │ 1 │ 1 2   (2)
   (6)   7 6 4 4 3 3 │ 2 │ 6   (1)
   (1)   1 │ 3 │ 5 5   (2)
   (2)   9 1 │ 4 │ 1 3   (2)
   (1)   1 │ 5 │ 0 0 2 8 4   (5)
   (3)   4 2 0 │ 6 │ 5 5 6 6   (5)
   (0)   │ 7 │ 3 4   (2)
                                   │ 8 │   (0)
                                   │ 9 │ 2 2 8 8   (4)
                                   │ 10 │   (0)
                                   │ 11 │ 1 3 3 4   (4)
                                   │ 12 │ 0 1   (2)

   (i) Find the median and quartiles for each month.

   (ii) Draw two box-and-whisker plots on the same diagram.

   (iii) Compare the distributions.

10   The table below gives the lengths, in minutes, of 50 telephone calls from a school office.

Length of call (x minutes)	$0 < x \leqslant 2$	$2 < x \leqslant 5$	$5 < x \leqslant 8$	$8 < x \leqslant 15$
Frequency	10	23	11	6

   (i) Draw a cumulative frequency graph to illustrate the data.

   (ii) Use your cumulative frequency graph to estimate the median and quartiles.

The shortest call was 1 minute and the longest call was 14 minutes.

   (iii) Draw a box-and-whisker plot and comment on the distribution.

11 The time, in minutes, that patients waited to be seen by a doctor was recorded for two different surgeries. The box-and-whisker plots show the results.

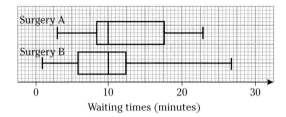

Waiting times (minutes)

Compare the waiting times at the two surgeries.

12 The employees in a firm were asked to record the distance, in km, they travelled by car in a given week. The distances, to the nearest km, are shown below.

67 76 85 42 93 48 93 46 52 72 77 53 41
48 86 78 56 80 70 70 66 62 54 85 60 58
43 58 74 44 52 74 52 82 78 47 66 50 67
87 78 86 94 63 72 63 44 47 57 68 81

(i) Construct a stem-and-leaf diagram to represent these data.

(ii) Find the median and the quartiles.

(iii) Draw a box-and-whisker plot to represent these data.

(iv) Give one advantage of using

　　(a)　a stem-and-leaf diagram,

　　(b)　a box-and-whisker plot.

to illustrate data such as those given above.

13 In a test on the protein quality of a new strain of corn, a farmer fed 20 newborn chicks with the new corn and observed how much weight they gained after three weeks. The results are given below.

Weight gain (grams): 360, 445, 403, 376, 434, 402, 397, 425, 407, 369,
　　　　　　　　462, 399, 427, 420, 410, 391, 430, 369, 410, 397

(i) Find the median and quartiles.

The farmer also fed a control group of 20 newborn chicks on the standard strain of corn he had previously used and recorded their weight gains after three weeks.

The lowest weight gain was 321 grams and the highest weight gain was 423 grams.

The median was 368.5 grams, the lower quartile was 353 grams and the upper quartile was 383 grams.

(ii) Draw two box-and-whisker plots on the same diagram showing the weight gains of the chicks fed the new strain of corn and the weight gains of the control group fed the standard strain of corn.

(iii) Use the box-and-whisker plots to compare the two distributions.

14 The back-to-back stem-and-leaf diagram below shows the time, in seconds, for the breathing rate to return to normal for a group of regular gym users and a group of people who do not exercise regularly.

> **Key:** 8 | 4 | 3 means 48 seconds for those who do not exercise regularly and 43 seconds for the regular gym users.

People who do not exercise regularly		Regular gym users	
	2	7 9	(2)
	3	0 2 3 6	(4)
(1) 8	4	3 5 6 8 9 9	(6)
(2) 9 4	5	1 3 5 7 7	(5)
(3) 7 1 0	6	0 2 4 4	(4)
(5) 8 7 4 2 1	7	3 5 8	(3)
(5) 9 9 9 5 3	8	1 1 5	(3)
(3) 6 4 0	9	2 6	(2)
(4) 9 7 7 4	10		
(2) 8 5	11		
(2) 1 0	12		
(1) 3	13		

(i) Find the median and quartiles for the regular gym users.

(ii) Find the median and quartiles for the people who do not exercise regularly.

(iii) Draw two box-and-whisker plots on the same diagram to represent the data.

(iv) Compare the times taken by the groups to return to normal breathing rate.

CHOOSING MEASURES AND DIAGRAMS

When representing data you need to consider the ways that would be most appropriate, for example:

- What type of diagram should you draw?
- Which average best represents the data?
- Which measure of spread is most informative?

The tables below give some of the advantages and disadvantages of using different measures and diagrams.

Measures of central tendency (averages)

	Advantages	Disadvantages
Mode	It is useful when the most popular category is needed, for example clothes or shoe sizes.	It is not useful in very small data sets or when there are more than two modes. There may not be a mode. It may not be representative, for example it could be the lowest value. The modal class depends on the grouping of the data. It is not useful for further analysis.

	Advantages	**Disadvantages**
Median	It is not affected by extreme values. It can be found as soon as a middle value is known, such as the distribution of times of runners in a race.	It does not use the whole data set. It is not useful for further analysis.
Mean	It is calculated using all the data and so represents every item. It is calculated using a mathematical formula, so calculators can be programmed to find it. It is extremely useful for further analysis.	It can be unduly affected by one or two extreme values.

Measures of variability (spread)

	Advantages	**Disadvantages**
Range	It is easy to calculate. It represents the complete spread of data.	It is affected by extreme values.
Interquartile range	It is not unduly influenced by extreme values. It can be used to investigate extreme values.	It depends only on particular values when the data are ranked.
Standard deviation	It is calculated using all the data and so represents every item. It is calculated using a mathematical formula, so calculators can be programmed to find it. It is very useful for further analysis. It is useful in comparing two sets of data, for example by showing which is more consistent.	It can be unduly affected by one or two extreme values. For a single set of data, its value is difficult to interpret.

Diagrammatic representation

	Advantages	**Disadvantages**
Bar chart	It shows the mode. Different sets of data can be compared using comparative bar charts.	It is only useful for qualitative data.
Pie chart	It shows the proportions of each quantity.	It has limited use with quantitative data. It does not show frequencies.
Vertical line diagram	It shows the mode clearly. It gives an idea of the shape of the distribution.	It is only useful for illustrating a small number of values.

	Advantages	Disadvantages
Stem-and-leaf diagram	It shows all the original data. It shows the shape of the distribution. The mode, median and quartiles can be found from the diagram. It is useful for comparing two sets of data.	It is not suitable for large amounts of data.
Histogram	It can represent groups of different widths. It shows whether the distribution is symmetrical or skew. The mean and standard deviation can be estimated from the histogram.	The visual impact can be altered by choosing different groups. Two distributions cannot be shown on the same diagram.
Cumulative frequency graphs	The median and quartiles can be estimated from the graph. Sets of data can be compared by drawing graphs on the same diagram.	The visual impact can be altered by using different scales.
Box-and-whisker plot	It is easy to see whether the distribution is symmetrical or whether there is a tail to the left or right. It can be used to investigate extreme values (outliers). It is easy to see the range and interquartile range. You can compare two or more sets of data by drawing plots on the same diagram.	It does not show frequencies.

Example 1.34

The times in minutes for seven students to become proficient at a new computer game were measured. The results are shown below.

$$15 \qquad 10 \qquad 48 \qquad 10 \qquad 19 \qquad 14 \qquad 16$$

(i) Find the mean and standard deviation of these times.

(ii) State which of the mean, median or mode you consider would be most appropriate to use as a measure of central tendency to represent the data in this case.

(iii) For each of the two measures of average you did not choose in part (ii), give a reason why you consider it inappropriate.

Cambridge Paper 62 Q1 J10

(i) Using the statistical functions on the calculator:
$\bar{x} = 18.85\ldots = 18.9$ (3 s.f.)
s.d. $= 12.26\ldots = 12.3$ (3 s.f.)

(ii) The median would be most appropriate.

(iii) The mode (10) is inappropriate because it is the lowest value.

The mean is inappropriate because it is affected by the outlier of 48.

Summary

Bar charts

There must be gaps between the bars.
The mode is given by the highest bar.

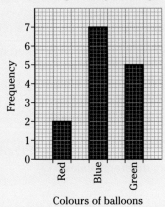

Colours of balloons

Pie charts

Pie charts show proportions in the different categories.

Vertical line graphs

The height of the line gives the frequency.
The mode is given by the highest line.

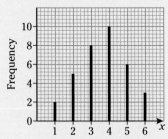

Stem-and-leaf diagrams (stemplot)

A key is essential.
Equal intervals must be chosen such as

   10–19, 20–29, 30–39, 40–49, 50–59.

Key: 2 | 7 means 27

Stem	Leaf
1	0 4 5 9
2	2 2 3 5 6 7 7
3	1 2 2 7 8
4	3 3 4 6
5	2 3 7

Histogram (grouped data)

$$\text{Frequency density} = \frac{\text{frequency}}{\text{interval width}}$$

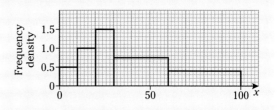

Area of bar = frequency in that interval

Total area = total frequency

There are no gaps between the bars.

Interval width = upper class boundary − lower class boundary

The modal class is represented by the highest bar.

Mean and standard deviation

For raw data:

$$\bar{x} = \frac{\sum x}{n} \qquad \text{standard deviation} = \sqrt{\frac{\sum(x - \bar{x})^2}{n}} = \sqrt{\frac{\sum x^2}{n} - \bar{x}^2}$$

For data in a frequency table:

Use the calculation versions.

$$\bar{x} = \frac{\sum xf}{\sum f} \qquad \text{standard deviation} = \sqrt{\frac{\sum(x - \bar{x})^2 f}{\sum f}} = \sqrt{\frac{\sum x^2 f}{\sum f} - \bar{x}^2}$$

When data are **grouped**, use the **mid-interval value** to represent the interval, where

$$\text{mid-interval value} = \tfrac{1}{2}(\text{l.c.b.} + \text{u.c.b.})$$

Combining sets of data for x and y:

$$\text{mean} = \frac{\sum x + \sum y}{n_1 + n_2} \qquad \text{standard deviation} = \sqrt{\frac{\sum x^2 + \sum y^2}{n_1 + n_2} - (\text{mean})^2}$$

Calculating mean and standard deviation given $\sum(x - a)$ and $\sum(x - a)^2$

To find the mean $\bar{x}$:

Find the mean of $(x - a)$, then add a

To find the standard deviation:

Find the standard deviation of $(x - a)$; this is the same as the standard deviation of x.

Cumulative frequency

The cumulative frequency is the total frequency up to a particular value.

Grouped data can be illustrated by a **cumulative frequency graph**.

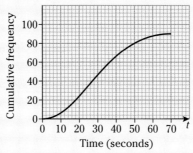

Plot cumulative frequencies against upper class boundaries.

Join the points with a smooth curve or with straight lines.

Median and quartiles

Median

For a set of n numbers arranged in ascending order:

– when n is odd, the median is the middle value

– when n is even, the median is the mean of the two middle values.

This is summarised by saying that the **median** is the $\frac{1}{2}(n + 1)^{\text{th}}$ value.

Quartiles

– The **lower quartile**, Q_1, is the median of all the values **before** the median.

– The **upper quartile**, Q_3, is the median of all the values **after** the median.

List the values in order and count through the data values to find the median and quartiles.

When estimating the median and quartiles from a cumulative frequency graph with a large total frequency, the following are usually used:

$$\text{median} = \tfrac{1}{2}n^{\text{th}} \text{ value, lower quartile} = \tfrac{1}{4}n^{\text{th}} \text{ value, upper quartile} = \tfrac{3}{4}n^{\text{th}} \text{ value.}$$

Ranges

Range = highest value − lowest value

Interquartile range (IQR) = upper quartile − lower quartile = $Q_3 - Q_1$

Box-and-whisker plot

– The ends of the whiskers are at the minimum and maximum values.

– The ends of the box are drawn at the lower quartile Q_1 and the upper quartile Q_3.

– The line in the box is drawn at the median Q_2.

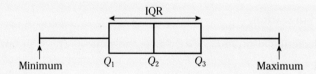

Shape of a distribution

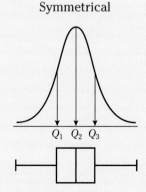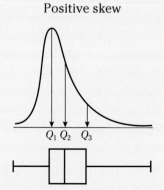

Mixed Exercise 1

1 The weights of 30 children in a class, to the nearest kg, were as follows.

   50   45   61   53   55   47   52   49   46   51
   60   52   54   47   57   59   42   46   51   53
   56   48   50   51   44   52   49   58   55   45

Construct a grouped frequency table for these data such that there are five equal class intervals with the first class having a lower boundary of 41.5 kg and the fifth class having an upper boundary of 61.5 kg.

Cambridge Paper 6 Q1 N06

2 Rachel measured the lengths in millimetres of some leaves on a tree. Her results are recorded below.

   32   35   45   37   38   44   33   39   36   45

Find the mean and standard deviation of the lengths of these leaves.

Cambridge Paper 6 Q1 N08

3 (i)

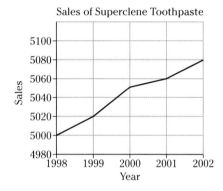

Sales of Superclene Toothpaste

The diagram represents the sales of Superclene toothpaste over the last few years. Give a reason why it is misleading.

(ii) The following data represent the daily ticket sales at a small theatre during three weeks.

   52, 73, 34, 85, 62, 79, 89, 50, 45, 83, 84, 91, 85, 84, 87, 44, 86, 41, 35, 73, 86

   (a)   Construct a stem-and-leaf diagram to illustrate the data.

   (b)   Use your diagram to find the median of the data.

Cambridge Paper 6 Q1 J03

4 In an experiment to estimate the value of π, Jon measured the circumference and diameter of several tins. He then divided the circumference by the diameter for each tin. His results are recorded below.

   3.05, 3.45, 3.19, 2.98, 2.85, 3.04, 3.28, 3.45, 4.87, 3.05

   (i) State the mode.

   (ii) Find the median.

   (iii) Calculate the mean.

   (iv) What percentage of Jon's results were higher than the true value of π?

5 The marks of 25 students in a test had a mean of 74 and a standard deviation of 8.

  (i) Find the total of the marks, $\sum x$.

  (ii) Show that $\sum x^2 = 138\,500$.

 It was later discovered that a mark of 86 had been entered incorrectly as 68.

  (iii) Calculate the mean and standard deviation of the corrected set of marks.

6 A study of the ages of car drivers in a certain country produced the results shown in the table.

Percentage of drivers in each age group

	Young	Middle-aged	Elderly
Males	20	30	50
Females	35	55	10

 Illustrate these results diagrammatically. Cambridge Paper 6 Q1 N05

7 The following table gives the marks, out of 75, in an examination taken by 243 students.

Marks	1–20	21–30	31–40	41–50	51–60	61–75
Frequency	24	38	72	60	34	15

  (i) Draw a histogram on graph paper to represent these results.

  (ii) Calculate estimates of the mean mark and the standard deviation.

8 Applicants for a job were asked to carry out a task to assess their practical skills. The times, in seconds, taken by 19 applicants were as follows:

   61, 229, 164, 76, 74, 49, 67, 86, 70, 82, 48, 74, 61, 59, 72, 81, 102, 61, 73

  (i) Find the mode, the median and the mean.

  (ii) Find the range, interquartile range and standard deviation.

  (iii) State, with a reason, which of the measures you have calculated you consider most appropriate

    (a) as a measure of central tendency,

    (b) as a measure of variability.

9 The pulse rates, in beats per minute, of a random sample of 15 small animals are shown in the following table.

   115 120 158 132 125
   104 142 160 145 104
   162 117 109 124 134

  (i) Draw a stem-and-leaf diagram to represent the data.

  (ii) Find the median and the quartiles.

  (iii) On graph paper, using a scale of 2 cm to represent 10 beats per minute, draw a box-and-whisker plot of the data. Cambridge Paper 6 Q5 N08

10 A computer can generate random numbers which are either 0 or 2. On a particular occasion, it generates a set of numbers which consists of 23 zeros and 17 twos. Find the mean and variance of this set of 40 numbers. Cambridge Paper 6 Q1 N03

11 The table below shows the frequency distribution of the masses of 52 women students at a college. Measurements have been recorded to the nearest kilogram.

Mass (kg)	40–44	45–49	50–54	55–59	60–64	65–69	70–74
Frequency	3	2	7	18	18	3	1

   (i) Construct a cumulative frequency table and draw a cumulative frequency graph.

  (ii) Estimate how many students weighed less than 57 kg.

 (iii) Estimate how many students weighed at least 61 kg.

 (iv) 20% weighed at least x kg. Estimate the value of x.

  (v) Estimate the median.

 (vi) Estimate the interquartile range.

12 The length of time, t minutes, taken to do the crossword in a certain newspaper was observed on 12 occasions. The results are summarised below.

$$\sum(t - 35) = -15 \qquad\qquad \sum(t - 35)^2 = 82.23$$

Calculate the mean and standard deviation for these times taken to do the crossword.

Cambridge Paper 6 Q1 J07

13 The diagram shows the cumulative frequency curves for data sets A, B and C.

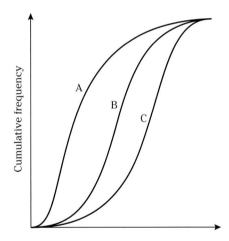

Below are the three frequency curves associated with sets A, B and C. Label each frequency curve with the appropriate letter.

(i)

(ii)

(iii)

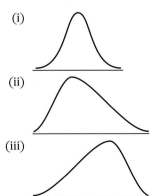

14 The weights in kilograms of two groups of 17-year-old males from country P and country Q are displayed in the following back-to-back stem-and-leaf diagram. In the third row of the diagram, ... 4 | 7 | 1 ... denotes weights of 74 kg for a male in country P and 71 kg for a male in country Q.

Country P		Country Q
	5	1 5
	6	2 3 4 8
9 8 7 6 4	7	1 3 4 5 6 7 7 8 8 9
8 8 6 6 5 3	8	2 3 6 7 7 8 8
9 7 7 6 5 5 5 4 2	9	0 2 2 4
5 4 4 3 1	10	4 5

(i) Find the median and quartile weights for country Q.

(ii) You are given that the lower quartile, median and upper quartile for country P are 84, 94 and 98 kg respectively. On a single diagram on graph paper, draw two box-and-whisker plots of the data.

(iii) Make two comments on the weights of the two groups.

Cambridge Paper 6 Q7 N02

15 A delivery company recorded the distance, x kilometres, driven each day by one of their drivers, over a period of 200 working days.

The mean distance driven each day by the driver was 300 km.

Given that $\sum(x - 300)^2 = 405\,000$, calculate the standard deviation.

16 Two hundred and fifty Army recruits have the following heights.

Heights (x cm)	$165 \leqslant x < 170$	$170 \leqslant x < 175$	$175 \leqslant x < 180$	$180 \leqslant x < 185$	$185 \leqslant x < 190$	$190 \leqslant x < 195$
Frequency	18	37	60	65	48	22

(a) Estimate the mean and standard deviation of the heights.

(b) (i) Draw a cumulative frequency curve to illustrate the data.

(ii) Use the curve to estimate the median height and the lower quartile height.

The tallest 40% of the recruits are to be formed into a special squad. For the members of the special squad, estimate

(iii) the median,

(iv) the upper quartile of the heights.

17 The times, to the nearest second, for 100 athletes to cover one lap of a running track were recorded and are shown in the table below.

Recorded time, x seconds	65–69	70–74	75–79	80–84	85–89	90–94	95–99
Frequency	0	8	20	25	31	10	6

(i) Draw a cumulative frequency graph and hence estimate the interquartile range.

To qualify for an athletics meeting, a runner needs to record a lap time of 78 seconds or under.

(ii) Estimate the number of athletes who qualified and the median time for these qualifiers.

18 The lengths of time, in minutes, to swim a certain distance by the members of a class of twelve 9-year-olds and the members of a class of eight 16-year-olds are shown below:

9-year-olds: 13.0 16.1 16.0 14.4 15.9 15.1 14.2 13.7 16.7 16.4 15.0 13.2
16-year-olds: 14.8 13.0 11.4 11.7 16.5 13.7 12.8 12.9

   (i) Draw a back-to-back stem-and-leaf diagram to represent the information above.

   (ii) A new pupil joined the 16-year-old class and swam the distance. The mean time for the class of nine pupils was now 13.6 minutes. Find the new pupil's time to swim the distance.

Cambridge Paper 6 Q4 J07

19

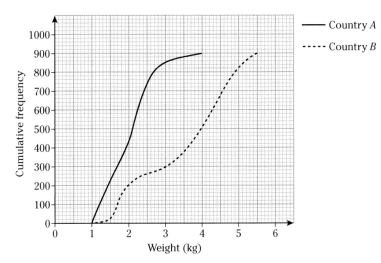

The birth weights of random samples of 900 babies born in country A and 900 babies born in country B are illustrated in the cumulative frequency graphs. Use suitable data from these graphs to compare the central tendency and spread of the birth weights of the two sets of babies.

Cambridge Paper 62 Q3 J10

20 During January the numbers of people entering a store during the first hour after opening were as follows.

Time after opening, x minutes	Frequency	Cumulative frequency
$0 < x \leqslant 10$	210	210
$10 < x \leqslant 20$	134	344
$20 < x \leqslant 30$	78	422
$30 < x \leqslant 40$	72	a
$40 < x \leqslant 60$	b	540

   (i) Find the values of a and b.

   (ii) Draw a cumulative frequency graph to represent the information. Take a scale of 2 cm for 10 minutes on the horizontal axis and 2 cm for 50 people on the vertical axis.

   (iii) Use your graph to estimate the median time after opening that people entered the store.

   (iv) Calculate estimates of the mean, m minutes, and standard deviation, s minutes, of the time after opening that people entered the store.

   (v) Use your graph to estimate the number of people entering the store between $(m - \frac{1}{2}s)$ and $(m + \frac{1}{2}s)$ minutes after opening.

Cambridge Paper 6 Q6 J09

2 Permutations and combinations

In this chapter you will learn about

- arrangements in a line
  - when items are distinct
  - when items are not distinct
  - when there are restrictions
  - when repetitions are allowed
- permutations of *r* items from *n* items
- combinations of *r* items from *n* items
  - when *n* items are distinct
  - when *n* items are not distinct
- methods to solve problems

ARRANGEMENTS IN A LINE

Arrangements of distinct items

How many different ways are there of arranging all four letters A, B, C, D in a line?

Suppose you have four spaces to be filled with the four letters A, B, C and D.

The first space can be filled in 4 ways with either A or B or C or D.

Now for <u>each</u> of these 4 choices, the second space can be filled in 3 ways.

So there are 4 × 3 ways of filling the first two spaces.

For <u>each</u> of these 12 choices, the third space can be filled in 2 ways.

So there are 4 × 3 × 2 ways of filling the first three spaces.

Then the fourth space can be filled in just one way.

So the number of ways of arranging the 4 letters in a line is

$$4 \times 3 \times 2 \times 1 = 24$$

Here are the 24 arrangements:

ABCD	ABDC	ACBD	ACDB	ADCB	ADBC
BCDA	BCAD	BDAC	BDCA	BACD	BADC
CDBA	CDAB	CABD	CADB	CBAD	CBDA
DABC	DACB	DBCA	DBAC	DCAB	DCBA

Notice that the letters being arranged are **distinct**. This means that they are all different from each other; no letter occurs more than once.

4 × 3 × 2 × 1 is called '4 factorial' and is written 4!

The number of ways of arranging 4 distinct items in a line is 4!

Directly on calculator

   24

This method can be extended to any number of distinct items arranged in a line.

For example
- the number of ways of arranging the letters P, Q, R, S, T, U, V in a line is
  $7 \times 6 \times 5 \times 4 \times 3 \times 2 \times 1 = 7! = 5040$
- the number of 5-digit numbers containing each of the digits 2, 3, 4, 5, 6 once only is
  $5 \times 4 \times 3 \times 2 \times 1 = 5! = 120$

The number of different arrangements of n distinct items is

$$n \times (n-1) \times (n-2) \times \ldots \times 3 \times 2 \times 1 = n!$$

n must be a positive integer

Example 2.1

Each of the letters of the word **CAMBRIDGE** is written on a card and the cards are placed in a line.

(i) How many different arrangements are there?

(ii) How many arrangements begin with **CAM**?

There are 9 letters and they are distinct (no repeat letters).

(i) Number of arrangements = $9! = 362\,880$

On calculator using factorials:

9 $\boxed{x!}$ $\boxed{=}$ $362\,880$

(ii) Place the first three letters **C A M**

There are now only 6 letters to be placed.

Number of arrangements = $6! = 720$

6 $\boxed{x!}$ $\boxed{=}$ 720

Some useful examples on factorials

$(n+1)! = (n+1) \times n!$

For example

$5! = 5 \times 4! = 5 \times 4 \times 3! = 5 \times 4 \times 3 \times 2! = 5 \times 4 \times 3 \times 2 \times 1 = 120$

$19! = 19 \times 18! = 19 \times 18 \times 17!$ and so on.

Calculations involving factorials

On calculator, using factorials

9 $\boxed{x!}$ $\boxed{\div}$ 6 $\boxed{x!}$ $\boxed{=}$ 504

$$\frac{9!}{6!} = \frac{9 \times 8 \times 7 \times \cancel{6} \times \cancel{5} \times \cancel{4} \times \cancel{3} \times \cancel{2} \times \cancel{1}}{\cancel{6} \times \cancel{5} \times \cancel{4} \times \cancel{3} \times \cancel{2} \times \cancel{1}} = 9 \times 8 \times 7 = 504$$

$$\frac{12!}{7!5!} = \frac{12 \times 11 \times 10 \times 9 \times 8 \times \cancel{7} \times \cancel{6} \times \cancel{5} \times \cancel{4} \times \cancel{3} \times \cancel{2} \times \cancel{1}}{\cancel{7} \times \cancel{6} \times \cancel{5} \times \cancel{4} \times \cancel{3} \times \cancel{2} \times \cancel{1} \times 5 \times 4 \times 3 \times 2 \times 1} = \frac{95\,040}{120} = 792$$

Note: If you calculate $\frac{12!}{7!5!}$ directly on your calculator using factorials, take care with the denominator. Here are two methods.

Method 1: Use brackets to multiply the terms in the denominator

$\boxed{12}$ $\boxed{x!}$ $\boxed{\div}$ $\boxed{(}$ $\boxed{7}$ $\boxed{x!}$ $\boxed{\times}$ $\boxed{5}$ $\boxed{x!}$ $\boxed{)}$ $\boxed{=}$ 792

Method 2: Divide by both the terms in the denominator

$\boxed{12}$ $\boxed{x!}$ $\boxed{\div}$ $\boxed{7}$ $\boxed{x!}$ $\boxed{\div}$ $\boxed{5}$ $\boxed{x!}$ $\boxed{=}$ 792

Take care here.

Arrangements when items are not distinct

What happens when the items to be arranged are not distinct, that is, some of the items are identical?

Suppose you have to find the number of different arrangements of the seven letters of the word EVEREST. These letters are not distinct as the 3 Es are identical. However, if you re-wrote them as E_1, E_2 and E_3 to make them distinct, then the number of ways to arrange the 7 distinct letters of the word $E_1VE_2RE_3ST$ would be 7!

Now look at these arrangements.

$E_1VE_2RE_3ST$	$E_2VE_1RE_3ST$	$E_3VE_1RE_2ST$
$E_1VE_3RE_2ST$	$E_2VE_3RE_1ST$	$E_3VE_2RE_1ST$

Notice that they are different with the subscripts, but identical when you remove the subscripts.

Since E_1, E_2, E_3 can be arranged amongst themselves in 3! = 6 ways, <u>each</u> arrangement of the 7 letters, such as EVEREST (shown above), VESTEER, SEVERET, ... and so on, would occur 3! times. So you need to divide 7! (the number of arrangements of 7 distinct letters) by 3! (the number of ways the 3 identical letters can be arranged amongst themselves).

So, the number of different arrangements

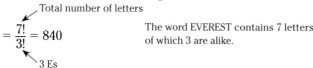

$$= \frac{7!}{3!} = 840$$

The word EVEREST contains 7 letters of which 3 are alike.

The number of different arrangements of n items of which p are alike is $\dfrac{n!}{p!}$.

What happens when several items are repeated?

The method can be extended by considering how many times each item appears. For example, the number of different arrangements of the letters of the word MINIMUM is

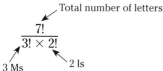

$$\frac{7!}{3! \times 2!}$$

The number of different arrangements of n items of which p of one type are alike, q of another type are alike, r of another type are alike, and so on, is $\dfrac{n!}{p! \times q! \times r! \times ...}$.

Example 2.2

Find the number of different arrangements using all ten letters of the word STATISTICS.

There are 10 letters to be arranged.

Consider the letters that are repeated.

> T occurs 3 times.
> S occurs 3 times.
> I occurs 2 times.

So the number of different arrangements

$$= \frac{10!}{3! \times 3! \times 2!} = 50\,400$$

3 Ts   3 Ss   2 Is

Arrangements when there are restrictions

What happens when there are restrictions on the arrangements that can be made?

Example 2.3

The word **ARGENTINA** includes the four consonants **R, G, N, T** and the three vowels **A, E, I**.

  (i) Find the number of different arrangements using all nine letters.

 (ii) How many of these arrangements have a consonant at the beginning, then a vowel, then another consonant, and so on alternately?

Cambridge Paper 6 Q1 N04

  (i) There are 9 letters.
A occurs twice and N occurs twice.
So the number of different arrangements

$$= \frac{9!}{2!2!} = 90\,720$$

2 As   2 Ns

 (ii) You want a consonant at the beginning (C), then a vowel
(V), then another consonant (C), and so on

      i.e. C V C V C V C V C

First place the consonants, noting that there are repeats.

Number of arrangements of the consonants R, G, N, N, T

$$= \frac{5!}{2!} = 60$$

2 Ns

Now place the vowels, noting that there are repeats.

The four vowels A, A, E, I can then slot into the four spaces between the consonants.

Number of arrangements of the vowels A, A, E, I

$$= \frac{4!}{2!} = 12$$

2 As

So the number of arrangements with consonants and vowels placed alternately

$$= \frac{5!}{2!} \times \frac{4!}{2!} = 60 \times 12 = 720$$

What happens when particular items have to be together, or must be separated?

Example 2.4

Issam has 11 different CDs, of which 6 are pop music, 3 are jazz and 2 are classical.

How many different arrangements of all 11 CDs on a shelf are there if the jazz CDs are all next to each other?

Cambridge Paper 6 Q3(i) J08

Since the CDs are all different, the items to be arranged are distinct.

Call the jazz CDs J_1, J_2 and J_3 and place them together.

Putting all the jazz CDs together, treat $\boxed{J_1 J_2 J_3}$ as one item.

So you now have 9 'items', which are

$$P_1, P_2, P_3, P_4, P_5, P_6, \boxed{J_1 J_2 J_3}, C_1, C_2$$

These 9 items can be arranged in 9! ways.

But $\boxed{J_1 J_2 J_3}$ can be arranged amongst themselves in 3! ways.

So the number of arrangements with the jazz CDs together

$$= 9! \times 3! = 2\,177\,280$$

Example 2.5

The eight sopranos in a choir are asked to stand in a line, but Ruby and Grace refuse to stand next to each other. How many different arrangements can there be?

Method 1

First find the number of arrangements with Ruby and Grace <u>together</u> and subtract this from the number of arrangements when there are no restrictions.

As in Example 2.4, treat $\boxed{RG}$ as one 'item'.

Together with the 6 other sopranos there are now 7 'items' to arrange and this can be done in 7! ways.

But Ruby and Grace can be arranged in 2! ways, i.e $\boxed{RG}$ and $\boxed{GR}$

So there are $7! \times 2!$ arrangements with Ruby and Grace <u>together</u>.

The number of arrangements of the 8 sopranos without any restrictions is 8!

So, the number of ways with Ruby and Grace <u>separated</u>
= number of ways of arranging 8 people
    – number of ways with Ruby and Grace together
$= 8! - 7! \times 2$
$= 30\,240$

You could tackle this problem in another way as follows.

Method 2

Without Ruby or Grace, the 6 remaining sopranos can be arranged in 6! ways.

This leaves 7 possible spaces for one of Ruby or Grace to stand in. She could stand in any of these 7 spaces, marked * below.

$$* \, S_1 \, * \, S_2 \, * \, S_3 \, * \, S_4 \, * \, S_5 \, * \, S_6 \, *$$

Once Ruby or Grace is placed, there are 6 possible spaces for the other one.

So the number of ways with Ruby and Grace separated
$= 6! \times 7 \times 6$
$= 30\,240$, as before.

Challenge:
Show, using factorials, that $8! - 7! \times 2 = 6! \times 7 \times 6$

Arrangements when repetitions are allowed

What happens when items may be repeated in the arrangement?

Example 2.6

How many 5-digit **odd** numbers can be made with the digits 2, 3, 6, 7, 8
 (i)   if repetitions are not allowed, for example, 63 287,
 (ii) if repetitions are allowed, for example, 88 663?

Since the number is odd, it must end in a 3 or a 7.

 (i) If repetitions are not allowed:

   There are 2 choices for the last digit.

   The remaining 4 digits can be arranged in 4! ways.

   So, if repetitions are not allowed, there are $2 \times 4! = 48$ possible odd numbers.

 (ii) If repetitions are allowed:

   There are 5 choices for each of the first 4 digits, but only 2 choices for the last digit.

   So, if repetitions are allowed, there are

   $5 \times 5 \times 5 \times 5 \times 2 = 5^4 \times 2 = 1250$ possible odd numbers.

Example 2.7

 (i)   Safebank requires its customers to use a four-digit PIN to access their account. Customers can choose any set of 4 digits from 0, 1, 2, ..., 9 and digits may be repeated. How many possible four-digit PINs are there?
 (ii) Smartbank requires its customers to use a password consisting of four lower-case letters. Repetitions are allowed. How many possible passwords are there?
 (iii) Excelbank requires its customers to use a pass-code consisting of four letters followed by four digits. Repetitions are allowed. How many possible pass-codes are there?

 (i)   There are 10 choices for each of the four digits, so number of possible
       PINs $= 10 \times 10 \times 10 \times 10 = 10^4 = 10\,000$

 (ii) There are 26 choices for each letter, so number of possible passwords
       $= 26 \times 26 \times 26 \times 26 = 26^4 = 456\,976$

 (iii) Number of possible pass-codes
       $= 26^4 \times 10^4 = 4\,569\,760\,000$

Exercise 2a

Section A – Practice with factorials

1 Evaluate the following without using a calculator, then check on your calculator using the $\boxed{x!}$ button.

(i) $7!$

(ii) $\dfrac{10!}{7!}$

(iii) $\dfrac{6!}{3!2!}$

(iv) $\dfrac{8!}{4!3!2!}$

2 Evaluate using the factorial button on your calculator.

(i) $\dfrac{8!}{5!} \times \dfrac{4!}{2!}$

(ii) $\dfrac{10!}{(5!)^2}$

3 Re-write using factorial notation.

(i) $5 \times 4 \times 3 \times 2 \times 1$ (ii) $5 \times 4 \times 3$

(iii) $21 \times 20 \times 19 \times 18$ (iv) $n(n-1)(n-2)$

(v) $\dfrac{12 \times 11 \times 10 \times 9}{5 \times 4 \times 3}$

Section B – Arrangements in a line

1 Find the number of different arrangements of

(i) 6 different-coloured toy engines on a track,

(ii) 7 people to be seated in a row on a bench,

(iii) 9 paintings to be hung in a row in a gallery.

2 Find the number of ways to arrange the letters of each of the following words.

(i) SPIDER   (ii) APRIL

(iii) SANDWICH

3 How many 4-digit numbers can be made with the digits 1, 2, 3, 4

(i) if repetitions are not allowed,

(ii) if repetitions are allowed?

4 The digits 4, 5, 7 and 9 are written on cards, as shown.

$\boxed{4}$ $\boxed{5}$ $\boxed{7}$ $\boxed{9}$

Mari places all the cards in a row to make a 4-digit number.

(i) How many even numbers can be made?

(ii) How many numbers can be made in which 5 and 7 are next to each other?

5 Find the number of ways to arrange the letters of each of the following words:

(i) PENGUIN

(ii) BLACKBERRY

(iii) MATHEMATICAL

6 Find the number of ways in which all eight letters of the word ADVANCED can be arranged if the arrangement must begin and end with an A.

7 Find the number of ways in which all eight letters of the word NEEDLESS can be arranged:

(i) if there are no restrictions,

(ii) if the arrangement must end with N,

(iii) if the three letters E must be placed next to each other,

(iv) if the two letters S must not be placed next to each other,

(v) if the three letters E must be placed together and the two letters S must not be placed together.

8 $\boxed{1}$ $\boxed{3}$ $\boxed{3}$ $\boxed{8}$ $\boxed{8}$ $\boxed{8}$

The above cards are placed in a line to form a 6-digit number. How many numbers can be made:

(i) if there are no restrictions,

(ii) if the number ends in 3,

(iii) if the number is odd?

9 Eloise is on a diet. Each day she can choose just one item from yoghurt, coffee or cereal for her breakfast. In how many ways can Eloise arrange her breakfast over 5 days?

10 A supermarket uses a five-digit number as a security code on its staff entrance. Digits may be repeated. How many different codes are possible when

(i) any of the ten digits from 0 to 9 may be used,

(ii) the zero cannot be used?

11 Five boys and four girls sit on a bench. In how many ways can they be seated if no two boys sit next to each other?

12 Matilda has eight different textbooks on a bookshelf. Two of them are mathematics books. In how many ways can the books be arranged if the two mathematics books must be placed:

(i) together,

(ii) with one at each end?

13 Three girls and seven boys stand in a line. Calculate the number of different arrangements if:

(i) the two youngest pupils are separated,

(ii) all three girls stand together.

14 There are 10 seats in a row in a theatre. In how many ways can 5 couples be seated in a row if each couple sits together?

15 Find how many different arrangements there are of the eleven letters of the word PROBABILITY if the two letters B are at the beginning and the two letters I are at the end.

16 Four boys and six girls are to be seated on a bench for a photograph. Calculate the number of different arrangements if:

(i) there is to be a boy at each end,

(ii) there is to be a girl at each end.

17 Four identical tins of peaches and six identical tins of pears are arranged in a row on a shelf. Calculate the number of different arrangements if the tin at each end contains the same type of fruit.

18 In how many ways can 5 boys and 3 girls stand in a straight line:

(i) if there are no restrictions,

(ii) if the boys stand next to each other?

Cambridge Paper 6 Q6(b) N03

19 Three identical yellow balloons, two identical red balloons and two identical blue balloons are strung in a row to celebrate Shema's birthday. Calculate the number of arrangements if:

(i) the balloon at each end is the same colour,

(ii) the yellow balloons are next to each other and the blue balloons are not next to each other.

20 Find how many arrangements there are of the nine letters in the words GOLD MEDAL

(i) if there are no restrictions in the order of the letters,

(ii) if the two letters D come first and the two letters L come last.

Cambridge Paper 6 Q7(b) J05

PERMUTATIONS OF r ITEMS FROM n ITEMS

You have seen that the number of arrangements in a line of n distinct items is $n!$

So, for example:

the number of arrangements of 3 distinct letters in 3 spaces is 3!

the number of arrangements of 8 distinct letters in 8 spaces is 8!

What happens if you have more letters than spaces?

Suppose you have 7 letters A, B, C, D, E, F, G and just 3 spaces to fill.

The first space can be filled in 7 ways.

The second space can then be filled in 6 ways.

The third space can then be filled in 5 ways.

So, number of ways of arranging 3 letters from 7 letters

$= 7 \times 6 \times 5 = 210$

The different arrangements of the letters are called **permutations**.

The number of permutations of 3 items taken from 7 items is written $_7P_3$.

You will be able to find $_7P_3$ directly on your calculator:

⑦ $\boxed{_nP_r}$ ③ ⑤ $=$ 210

Note that $7 \times 6 \times 5$ could be written $\dfrac{7 \times 6 \times 5 \times 4 \times 3 \times 2 \times 1}{4 \times 3 \times 2 \times 1} = \dfrac{7!}{4!} = \dfrac{7!}{(7 - 3)!}$

So $_7P_3 = \dfrac{7!}{(7 - 3)!}$

The **order** in which the letters are arranged is important since, for example, ABC is a different permutation from ACB.

The number of **permutations**, or **ordered arrangements**, of r items taken from n distinct items is

$_nP_r = \dfrac{n!}{(n - r)!}$ $_nP_r$ can also be written ^nP_r

In permutations, the **order** of the selection **matters**.

Special case

Suppose you want to use $_nP_r$ to find the number of permutations of 7 letters taken from 7 letters (in other words, you take them all).

You have $_7P_7 = \dfrac{7!}{(7 - 7)!} = \dfrac{7!}{0!}$

But you know that the number of ways of arranging 7 distinct items is 7! In order that both expressions give the same value, 0! is <u>defined</u> to be 1.

You now have $\dfrac{7!}{0!} = \dfrac{7!}{1} = 7!$ Try ⓪ $\boxed{x!}$ $=$ on your calculator.

By definition, $0! = 1$

Example 2.8

Find how many numbers bigger than 30 000 but smaller than 40 000 can be formed from the digits 2, 3, 4, 5, 6, 7, 8 if no digit is repeated and the number must be a multiple of 5.

The number must have 5 digits.

It must start with 3 and end with 5, so fix these two digits

$\boxed{3}$ $\boxed{*}$ $\boxed{*}$ $\boxed{*}$ $\boxed{5}$

There are now three spaces to fill and the digits must be taken from the five digits 2, 4, 6, 7, 8.

Number of ways to fill the three remaining places

$= {_5P_3} = \dfrac{5!}{(5 - 3)!} = \dfrac{5!}{2!} = 60$ Directly on calculator:

⑤ $\boxed{_nP_r}$ ③ $=$ 60

So 60 different numbers can be made which satisfy the conditions.

Example 2.9

A security code consists of 4 letters chosen from A, B, C, D, E, F, G followed by 3 digits chosen from 0, 1, 2, 3, 4, 5.

Examples are BCDG102 (without repetitions) and CCDD225 (with repetitions).

Show that more than five times as many codes can be made when repetitions are allowed than when repetitions are not allowed.

There are 7 letters and 6 digits.

When repetitions are not allowed:

Number of arrangements of 4 letters from 7 letters

$$= {}_7P_4 = \frac{7!}{(7-4)!} = \frac{7!}{3!} = 840$$

Number of arrangements of 3 digits from 6 digits

$$= {}_6P_3 = \frac{6!}{(6-3)!} = \frac{6!}{3!} = 120$$

Number of possible codes (repetitions not allowed)

$$= 840 \times 120 = 100\,800$$

When repetitions are allowed:

There are 7 choices for each of the 4 letters, so number of choices $= 7^4$.

There are 6 choices for each of the 3 digits, so number of choices $= 6^3$.

Number of possible codes (repetitions allowed)

$$= 7^4 \times 6^3$$

$$= 518\,616 > 5 \times 100\,800$$

So more than five times as many codes can be made when repetitions are allowed.

Exercise 2b

1 Evaluate the following without using a calculator. Then check using the ${}_nP_r$ button on the calculator.
   (i) ${}_9P_6$ (ii) ${}_6P_2$
   (iii) ${}_{10}P_3$ (iv) ${}_9P_9$

2 From a class of 20 pupils, 3 pupils are going to be chosen to be sports officials. The first pupil chosen will be the swimming captain, the second pupil the athletics captain and the third pupil the tennis captain. In how many different ways can the officials be chosen?

3 There are 12 contestants in a singing competition. In how many ways can the first, second, third and fourth prizes be awarded?

4 In a particular minibus there are 16 seats for passengers. How many possible seating arrangements are there for 5 passengers?

5 How many even numbers between 6000 and 7000 can be formed using the digits 1, 2, 3, 6, 7, 9, if no digit is repeated?

6 A security code consists of 3 digits chosen from 4, 5, 6, 7, 8 followed by 2 letters chosen from P, Q, R, S, T, for example, 674TQ. How many different codes are possible

   (i) if repetitions are not allowed,

   (ii) if repetitions are allowed?

7 Rory is playing a game in which he has to place coloured pegs into holes in a board. He has 6 identical red pegs and the board has 10 holes. How many different arrangements are there for placing the 6 pegs and leaving 4 empty holes?

8 If repetitions are not allowed, how many numbers can be formed with the digits 3, 4, 5, 6, 7

(i) using three of the digits,

(ii) using one or more of the digits?

9 There are 10 seats in the front row at a theatre. Six people are shown to this row. In how many different ways can they be seated if

(i) there are no restrictions,

(ii) two particular people in the group must sit next to each other?

COMBINATIONS OF r ITEMS FROM n ITEMS

A **combination** is a selection of some items where the order of the selected items **does not matter**.

Consider this situation: the organisers of a local charity are raising money by selling tickets for a draw. To take part you fill in a ticket by selecting five numbers from the set of 30 numbers: 1, 2, 3, ..., 29, 30. The organisers make the draw by using a calculator to randomly generate five whole numbers between 1 and 30 inclusive. If these numbers are the same as the numbers on your ticket, you win a prize. The order in which the five numbers are arranged does not matter. How many different selections are there?

If the order mattered, the number of arrangements would be

$$_{30}P_5 = 30 \times 29 \times 28 \times 27 \times 26 = 17\,100\,720$$

But the order does not matter. So, for example, the following would all count as being the same:

8, 13, 21, 2, 17   21, 2, 13, 8, 17   13, 17, 2, 8, 21 …. and so on.

Now for *each* selection of 5 numbers there are 5! arrangements that would count as being the same.

So, the number of different selections of 5 from 30 when the order does not matter

$$= \frac{_{30}P_5}{5!} = 142\,506.$$

This is known as the number of **combinations** of 5 from 30 and is denoted by $\binom{30}{5}$ where

$$\binom{30}{5} = \frac{30!}{5! \times 25!} = \frac{30 \times 29 \times 28 \times 27 \times 26}{5 \times 4 \times 3 \times 2 \times 1}$$

Cancel 25! on the top and bottom of the fraction.

Say "30 choose 5".

Note that $\binom{30}{5}$ can also be written as $_{30}C_5$ or $^{30}C_5$ and you can find it directly on your calculator by keying in $\boxed{3}\ \boxed{0}\ \boxed{_nC_r}\ \boxed{5}\ \boxed{=}$

In general:

The number of **combinations** of r items chosen from n distinct items is given by

$$\binom{n}{r} = {_nC_r} = \frac{n!}{r!(n-r)!}$$

$_nC_r$ can also be written ^nC_r

n choose r

In **combinations**, the **order** of the selection **does not matter**.

Example 2.10

A team of 3 is to be chosen from 10 athletes. How many different teams could be chosen?

As the athletes are being chosen just to be in the team, rather than to take on specific roles, the order in which they are chosen does not matter, so use combinations.

Number of ways of choosing the team

$$= \binom{10}{3} = \frac{10!}{3!(10-3)!} = \frac{10!}{3!7!} = 120$$

10 choose 3

Directly on your calculator:

$\boxed{1}\ \boxed{0}\ \boxed{{}_nC_r}\ \boxed{3}\ \boxed{=}\ 120$

Example 2.11

Without using a calculator, evaluate $\binom{12}{9}$ and $\binom{12}{3}$.

$$\binom{12}{9} = \frac{12!}{9! \times 3!} = \frac{12 \times 11 \times 10 \times 9!}{9! \times 3 \times 2 \times 1}$$

Cancel 9!

$$= \frac{12 \times 11 \times 10}{3 \times 2 \times 1}$$
$$= 2 \times 11 \times 10$$
$$= 220$$

Now check ${}_nC_r$ on the calculator:

$\boxed{12}\ \boxed{{}_nC_r}\ \boxed{9}\ \boxed{=}$

Similarly

$$\binom{12}{3} = \frac{12!}{3! \times 9!} = \frac{12 \times 11 \times 10 \times 9!}{3 \times 2 \times 1 \times 9!} = \frac{12 \times 11 \times 10}{3 \times 2 \times 1} = 220$$

Notice that $\binom{12}{9} = \binom{12}{3}$

This is because choosing a group of 9 from 12 automatically leaves you with a group of 3.

Generally, $\binom{n}{r} = \binom{n}{n-r}$.

So the number of ways of choosing 9 from 12 is the same as the number of ways of choosing 3 from 12.

Example 2.12

Issam has 11 different CDs of which 6 are pop music, 3 are jazz and 2 are classical. Issam makes a selection of 2 pop music CDs, 2 jazz CDs and 1 classical CD.

How many different possible selections can be made?

Cambridge Paper 6 Q3(ii) part J08

Number of ways to choose 2 pop music CDs from 6 pop music CDs $= \binom{6}{2}$

Number of ways to choose 2 jazz CDs from 3 jazz CDs $= \binom{3}{2}$

Number of ways to choose 1 classical CD from 2 classical CDs $= \binom{2}{1}$

So number of possible selections

$$= \binom{6}{2} \times \binom{3}{2} \times \binom{2}{1} = 15 \times 3 \times 2 = 90$$

Example 2.13

A collection of 18 books contains one Harry Potter book. Linda is going to choose 6 of these books to take on holiday.

  (i) In how many ways can she choose 6 books?

  (ii) How many of these choices will include the Harry Potter book?

Cambridge Paper 6 Q6(a) N03

  (i) Number of ways to choose 6 books from 18 books

$$= \binom{18}{6} = \frac{18!}{6!12!} = 18\,564$$

  (ii) If the Harry Potter book is included, Linda has to choose the other 5 books from 17 books.

Number of ways including the Harry Potter book

$$= \binom{17}{5} = \frac{17!}{5!12!} = 6\,188$$

Example 2.14

A committee of 5 people is to be chosen from 6 men and 4 women. In how many ways can this be done:

  (i) if there must be 3 men and 2 women on the committee,

  (ii) if there must be more men than women on the committee,

  (iii) if there must be 3 men and 2 women, and one particular woman refuses to be on the committee with one particular man?

Cambridge Paper 6 Q5 J03

  (i) Number of ways to choose 3 men from 6 men $= \binom{6}{3}$

Number of ways to choose 2 women from 4 women $= \binom{4}{2}$

So, number of ways to choose committee

$$= \binom{6}{3} \times \binom{4}{2}$$
$$= 20 \times 6$$
$$= 120$$

  (ii) If there are more men than women on the committee there could be

	Number of ways
5 men, 0 women	$\binom{6}{5} = 6$
4 men, 1 woman	$\binom{6}{4} \times \binom{4}{1} = 15 \times 4 = 60$
3 men, 2 women	$\binom{6}{3} \times \binom{4}{2} = 20 \times 6 = 120$

Total number of ways $= 6 + 60 + 120 = 186$.

(iii)   Denoting the particular man by M and the particular woman by W, first consider the number of ways with <u>both</u> M and W on the committee.

You now need to choose two more men and one more woman.

Number of ways to choose 2 men from 5 men $= \binom{5}{2}$

Number of ways to choose 1 woman from 3 women $= \binom{3}{1}$

So number of ways with **both M and W** on the committee

$$= \binom{5}{2} \times \binom{3}{1} = 10 \times 3 = 30$$

Now subtract this from the number of ways of choosing the committee with no restrictions.

From part (i), total number of ways to choose committee $= 120$

So number of ways with **not both M and W** on the committee
$$= 120 - 30$$
$$= 90$$

Combination of *r* items from *n* items when the items are not distinct

What happens when the selections are from items that are not distinct?

Example 2.15

Three letters are selected at random from the letters of the word BIOLOGY.

Find the total number of selections.

*The answer is **not** $\binom{7}{3}$ as you might expect.*

Because there are two letters O, you need to find the number of selections with

   no letters O
   one letter O
   two letters O

and then add these together.

For example
   B, L and Y

Number of selections with no letter O

   $=$ number of ways to choose **three** letters from B, I, L, G, Y

   $= \binom{5}{3} = 10$

For example
   O, B and L

Number of selections with one letter O

   $=$ number of ways to choose **two** letters from B, I, L, G, Y

   $= \binom{5}{2} = 10$

For example
   O, O and B

Number of selections with two letters O

   $=$ number of ways to choose **one** letter from B, I, L, G, Y

   $= 5$

Therefore, total number of selections $= 10 + 10 + 5 = 25$

Exercise 2c

1 Evaluate each of the following without using a calculator, then check using the $_nC_r$ button on your calculator:

   (i) $\binom{11}{4}$ (ii) $\binom{7}{3}$ (iii) $\binom{8}{5}$ (iv) $\binom{10}{8}$

   (v) $\binom{9}{6}$ (vi) $\binom{9}{3}$ (vii) $\binom{12}{8}$ (viii) $\binom{12}{4}$

2 How many different teams of 5 people can be chosen, without regard to order, from a squad of 12 people?

3 A committee of four is to be chosen from a group of nine people which includes Mr Green, Mrs Green and Mr Brown. How many different selections are possible if:

   (i) there are no restrictions,

   (ii) Mr Brown must be on the committee,

   (iii) Mr Green and Mrs Green may not both be on the committee,

   (iv) Mr Green and Mrs Green must be on the committee but Mr Brown must not be on the committee?

4 Sadie has 6 blouses and 8 skirts. She is going on holiday and decides to pack 4 blouses and 4 skirts. How many different selections are possible?

5 A Spelling-Bee team of 5 students is to be chosen from a class of 12 boys and 9 girls. In how many ways can a team be chosen if the team consists of:

   (i) 3 boys and 2 girls,

   (ii) 3 girls and 2 boys,

   (iii) at least 3 girls?

6 How many different selections of five letters from the eleven letters of the word PROBABILITY contain both letters B and no vowels?

7 In how many ways can a group of 14 people eating at a restaurant be divided between 3 tables seating 6, 5 and 3?

8 A group of 12 guests at a wedding are to travel as passengers from the church to the reception. There are 3 cars: black, silver and blue. Each car holds 4 passengers. Find the number of ways in which the group may travel if Alice and Jack refuse to travel in the same car.

9 A football team consists of 3 players who play in a defence position, 3 players who play in a midfield position and 5 players who play in a forward position. Three players are chosen to collect a gold medal for the team. Find in how many ways this can be done

   (i) if the captain, who is a midfield player, must be included, together with one defence and one forward player,

   (ii) if exactly one forward player must be included, together with any two others.

   Cambridge Paper 6 Q7(a) part J05

10 Three letters are selected at random from the letters of the word PARABOLA. Find the total number of selections.

11 In a mixed pack of coloured light bulbs there are three red bulbs, one yellow bulb, one blue bulb and one green bulb. Four bulbs are selected at random from the pack. How many different selections are possible?

12 There are 20 teachers at a conference. Of these, 8 are maths teachers, 6 are history teachers, 4 are physics teachers and 2 are geography teachers.

   Four of the teachers are to be chosen at random to take part in a quiz. In how many different ways can the teachers be chosen:

   (i) if there is to be a teacher from each subject,

   (ii) if they all must teach the same subject,

   (iii) if there are to be at least two maths teachers?

13 A diagonal of a polygon is defined to be a line joining any two non-adjacent vertices.

   (i) Show that the number of diagonals in a 5-sided polygon is $\binom{5}{2} - 5$.

   (ii) How many diagonals are there in a 6-sided polygon?

   (iii) Show that the number of diagonals in an n-sided polygon is $\dfrac{n(n-3)}{2}$.

Summary

Arrangements in a line

The number of different arrangements of n distinct objects is

n must be a positive integer

$$n \times (n - 1) \times (n - 2) \times \ldots \times 3 \times 2 \times 1 = n!$$

The number of different arrangements of n items of which p are alike is $\dfrac{n!}{p!}$.

The number of different arrangements of n items of which p of one type are alike, q of another type are alike, r of another type are alike, and so on, is $\dfrac{n!}{p! \times q! \times r! \times \ldots}$.

By definition, $0! = 1$

Permutations and combinations

Permutations
The number of **permutations** of r items taken from n distinct items is
$$_nP_r = \frac{n}{(n - r)!}$$
In permutations, **order matters**.

Combinations
The number of **combinations** of r items taken from n distinct items is
$$_nC_r = \binom{n}{r} = \frac{n}{r!(n - r)!}$$
In combinations, **order does not matter**.

Mixed Exercise 2

1 (i) Find the number of ways in which all 12 letters of the word REFRIGERATOR can be arranged:

(a) if there are no restrictions,

(b) if the Rs must all be together.

(ii) How many different selections of four letters from the 12 letters of the word REFRIGERATOR contain no Rs and two Es?

Cambridge Paper 6 Q5 J07

2 A choir consists of 13 sopranos, 12 altos, 6 tenors and 7 basses. A group consisting of 10 sopranos, 9 altos, 4 tenors and 4 basses is to be chosen from the choir.

(i) In how many different ways can the group be chosen?

(ii) In how many ways can the 10 chosen sopranos be arranged in a line if the 6 tallest stand next to each other?

(iii) The 4 tenors and 4 basses stand in a single line with all the tenors next to each other and all the basses next to each other. How many possible arrangements are there if 3 of the tenors refuse to stand next to any of the basses?

Cambridge Paper 6 Q4 J09

3 Nine cards, each of a different colour, are to be arranged in a line.

(i) How many different arrangements of the 9 cards are possible?

The 9 cards include a pink card and a green card.

(ii) How many different arrangements do not have the pink card next to the green card?

Consider all possible choices of 3 cards from the 9 cards with the 3 cards being arranged in a line.

(iii) How many different arrangements in total of 3 cards are possible?

(iv) How many of the arrangements of the 3 cards in part (iii) contain the pink card?

(v) How many of the arrangements of 3 cards in part (iii) do not have the pink card next to the green card?

Cambridge Paper 62 Q7 J10

4 (i) Find the number of ways that a set of 10 different CDs can be shared between Dai and Evan if each receives an odd number of CDs.

(ii) A set of 9 DVDs consists of 3 different horror films and 6 different musicals. In how many ways can these be arranged on a shelf if the horror films are separated from each other?

5 (a) Find how many numbers between 5000 and 6000 can be formed from the digits 1, 2, 3, 4, 5 and 6

(i) if no digits are repeated,

(ii) if repeated digits are allowed.

(b) Find the number of ways of choosing a school team of 5 pupils from 6 boys and 8 girls

(i) if there are more girls than boys in the team,

(ii) if 3 of the boys are cousins and are either all in the team or all not in the team.

Cambridge Paper 61 Q5 N09

6 The digits of the number 1 223 678 can be rearranged to give many different 7-digit numbers.

Find how many different 7-digit numbers can be made if:

(i) there are no restrictions on the order of the digits,

(ii) the digits 1, 3, 7 (in any order) are next to each other,

(iii) these 7-digit numbers are even.

Cambridge Paper 6 Q5 J02

7 Four letters are selected at random from the letters of the word GEOGRAPHY. Find the total number of selections.

8

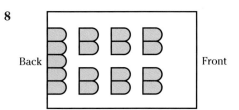

Back Front

The diagram shows the seating plan for passengers in a minibus, which has 17 seats arranged in 4 rows. The back row has 5 seats and the other 3 rows have 2 seats on each side. 11 passengers get on the minibus.

(i) How many possible seating arrangements are there for the 11 passengers?

(ii) How many possible seating arrangements are there if 5 particular people sit in the back row?

Of the 11 passengers, 5 are unmarried and the other 6 consist of 3 married couples.

(iii) In how many ways can 5 of the 11 passengers on the bus be chosen if there must be 2 married couples and 1 other person, who may or may not be married?

Cambridge Paper 6 Q4 J06

9 In a certain hotel, the lock on the door to each room can be opened by inserting a key card. The key card can be inserted only one way round. The card has a pattern of holes punched in it. The card has 4 columns and each column can have either 1 hole, 2 holes, 3 holes or 4 holes punched in it. Each column has 8 different positions for the holes. The diagram illustrates one particular key card with 3 holes punched in the first column, 3 in the second, 1 in the third and 2 in the fourth.

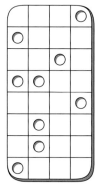

(i) Show that the number of different ways in which a column could have exactly 2 holes is 28.

(ii) Find how many different patterns of holes can be punched in a column.

(iii) How many different possible key cards are there?

Cambridge Paper 6 Q4 N02

3 Probability

In this chapter you will learn about

- the probability scale
- ways of estimating probabilities
- probability notation
- using arrangements, permutations and combinations to calculate probabilities
- the rule for combined events
- mutually exclusive events
- conditional probability
- independent events
- using tree diagrams

PROBABILITY

The **probability** of an event is a measure of the likelihood that it will happen. It is given on a numerical scale from 0 to 1 and the numbers representing probabilities can be written as decimals, fractions or percentages.

The two extremes on the probability scale are **impossibility** at one end and **certainty** at the other end.

A probability of 0 indicates that the event is **impossible**.

A probability of 1 (or 100%) indicates that the event is **certain** to happen.

All other events have a probability between 0 and 1.

For example:
- if a fair (unbiased) tetrahedral die with faces marked 1, 2, 3, 4 is thrown, the probability that it lands on 1 is $\frac{1}{4}$ or 0.25 or 25%
- if a fair coin is tossed, the probability that it shows heads is $\frac{1}{2}$ or 0.5 or 50%
- if you select a tennis ball from a box of **yellow** tennis balls
  - the probability that you will select a yellow ball is 1 (certain)
  - the probability that you will select a blue tennis ball is 0 (impossible).

These probabilities are shown on a probability scale below:

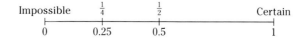

Experimental probability

When a drawing pin is dropped it lands in one of two positions: 'point-up' or 'point-down'.

point-up   point-down

What is the probability that the drawing pin will land 'point-up'?

To estimate the probability that the drawing pin will land 'point-up' you could drop the drawing pin a number of times and work out the proportion that land 'point-up'. This proportion is called the **relative frequency**, where

$$\text{relative frequency} = \frac{\text{number of 'points-up'}}{\text{total number of times the pin is dropped}}$$

But how many times should you drop the drawing pin: 10, 50, 100, 200, ...?

To investigate this, an experiment was carried out and the relative frequency of 'points-up' calculated after 10, 20, 30, 40, 50, ..., 100, 110, ..., 200 throws. The results are shown in the table.

Number of times pin dropped	Cumulative number of points-up	Relative frequency (2 d.p.)
10	3	$\frac{3}{10} = 0.3$
20	11	$\frac{11}{20} = 0.55$
30	16	$\frac{16}{30} = 0.53$
40	21	$\frac{21}{40} = 0.53$
50	28	$\frac{28}{50} = 0.56$
60	34	$\frac{34}{60} = 0.57$
70	40	$\frac{40}{70} = 0.57$
80	45	$\frac{45}{80} = 0.56$
90	48	$\frac{48}{90} = 0.53$
100	55	$\frac{55}{100} = 0.55$
110	62	$\frac{62}{110} = 0.56$
120	69	$\frac{69}{120} = 0.58$
130	74	$\frac{74}{130} = 0.57$
140	78	$\frac{78}{140} = 0.56$
150	86	$\frac{86}{150} = 0.57$
160	93	$\frac{93}{160} = 0.58$
170	101	$\frac{101}{170} = 0.59$
180	108	$\frac{108}{180} = 0.60$
190	115	$\frac{115}{190} = 0.61$
200	122	$\frac{122}{200} = 0.61$

The relative frequencies were then plotted on a graph, shown below.

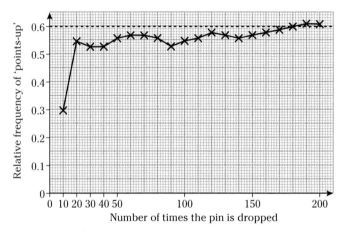

Notice that when the pin was dropped 10 times, the relative frequency was 0.3. However, when it was dropped more and more times, the relative frequency appears to be settling to a value around 0.6. This 'limiting' value is taken as an estimate of the probability that the pin will land 'point-up'.

In general, if an experiment is repeated n times under exactly the same conditions and a particular event occurs r times, then the relative frequency $\frac{r}{n}$ is an estimate of the probability of this event. This is known as experimental probability. Note that the accuracy of the estimate increases as n increases: the larger the value of n, the better the estimate.

Subjective probability

Suppose you want to estimate the probability that it will snow in Cambridge on Christmas Day or the likelihood that a house in a particular area of a city will be burgled. You could form a **subjective probability,** based on past experience, such as weather records or crime figures, or on expert opinion or other factors. This method is open to error as two people considering the same evidence may give different estimates of the probability. However, it is sometimes the only method available.

Theoretical probability

If you were asked the probability of obtaining a head when a fair coin is tossed, you would give the answer $\frac{1}{2}$ without bothering to toss a coin a large number of times and working out the relative frequency. Intuitively you use the definition of probability when outcomes are equally likely, where

$$\text{probability} = \frac{\text{number of successful outcomes}}{\text{total number of possible outcomes}}$$

Notation
It is useful to use some shorthand notation.

The set of all possible outcomes is the **possibility space**, S, and the number of outcomes in the possibility space is written $n(S)$.

The event A consists of one or more of the outcomes in S. The number of outcomes resulting in event A is written $n(A)$.

It can be helpful to draw a **Venn diagram**, where the possibility space S is denoted by a rectangle and the event A by a circle or oval drawn inside the rectangle.

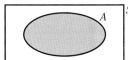

When outcomes are **equally likely**, the **probability** of event A is written P(A) where

$$P(A) = \frac{n(A)}{n(S)}$$
⟵ number of outcomes resulting in event A
⟵ total number of outcomes in the possibility space S

The **complement** of A is A' where A' is the event **A does not occur**.

Since an event either occurs or does not occur,

$$P(A) + P(A') = 1$$

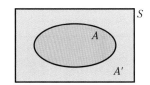

i.e.　　　$P(A') = 1 - P(A)$

The notation A' will be used in the examination, but note that an alternative notation is $\overline{A}$.

Consider these two examples:

- When an ordinary fair cubical die is thrown, the possibility space for the score on the die (i.e. the number on the uppermost face) consists of the numbers from 1 to 6.
  This can be written $S = \{1, 2, 3, 4, 5, 6\}$ and $n(S) = 6$.

  If event A is 'the score is lower than 3', then $A = \{1, 2\}$ and $n(A) = 2$.

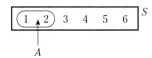

  So,　　$P(A) = \dfrac{n(A)}{n(S)} = \dfrac{2}{6} = \dfrac{1}{3}$

  Also, $P(A') = P(\text{the score is 3 or more}) = 1 - \dfrac{1}{3} = \dfrac{2}{3}$

• When two fair dice are thrown, there are 36 possible outcomes, so $n(S) = 36$. It is convenient to show these by dots on a possibility space diagram.

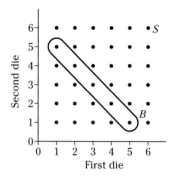

If event B is 'the sum of the scores is 6', then the outcomes resulting in B are $(1, 5), (2, 4), (3, 3), (4, 2), (5, 1)$. These are shown ringed in the diagram and you can see that $n(B) = 5$.

So, $P(B) = \dfrac{n(B)}{n(S)} = \dfrac{5}{36}$

Also, $P(B') = P(\text{sum of the scores is not 6}) = 1 - \dfrac{5}{36} = \dfrac{31}{36}$

Example 3.1

A box contains 20 counters numbered 1, 2, 3, ... up to 20. A counter is picked at random from the box. Find the probability that the number on the counter is

  (i)   a multiple of 5,

  (ii)   not a multiple of 5,

  (iii)   higher than 7.

$S = \{1, 2, 3, ..., 20\}$ so $n(S) = 20$.

  (i)   Let A be the event 'the number is a multiple of 5'.   Each outcome is
$A = \{5, 10, 15, 20\}$, so $n(A) = 4$.   equally likely to occur.

     $P(A) = \dfrac{n(A)}{n(S)} = \dfrac{4}{20} = 0.2$

  (ii) $P(\text{not a multiple of 5}) = P(A') = 1 - 0.2 = 0.8$ $P(A') = 1 - P(A)$

  (iii)   Let B be the event 'the number is higher than 7'.
$B = \{8, 9, ..., 20\}$ so $n(B) = 13$.   Do not include 7.

     $P(B) = \dfrac{n(B)}{n(S)} = \dfrac{13}{20} = 0.65$

Example 3.2

A fair five-sided spinner has sides numbered 1, 1, 2, 3, 3.

The spinner is spun twice.
Find the probability that the spinner will stop at 1 at least once.

When the spinner is spun twice there are 25 outcomes (shown by dots on the possibility space diagram), so $n(S) = 25$.

Each outcome is equally likely to occur.

Let A be the event 'the spinner will stop at 1 at least once'.

From the diagram you can see that $n(A) = 16$.

So $P(A) = \dfrac{16}{25} = 0.64$

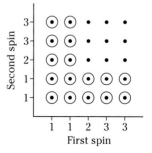

Often questions involve a set of playing cards.

An ordinary pack of playing cards consists of 52 cards, split equally into four suits:

Jack of hearts　　2 of diamonds　　Ace of spades　　9 of clubs

Diamonds and hearts are red suits; clubs and spades are black suits.

In each suit there are 13 cards:

Ace, 2, 3, 4, 5, 6, 7, 8, 9, 10, Jack, Queen, King

The Jack, Queen and King of any suit are called 'picture cards'.

Example 3.3

A card is dealt from a well-shuffled ordinary pack of 52 playing cards.

(i) Find the probability that the card is

   (a) the 4 of spades,　　 (b) the 4 of spades or any diamond.

(ii) The first card dealt is placed face-up on the table. It is the 3 of diamonds. What is the probability that the second card is from a red suit?

(i) When a card is dealt, it is equally likely to be any of the 52 cards, so $n(S) = 52$.

  (a) Let A be the event 'the card is the 4 of spades'. There is only one '4 of spades' in the pack, so $n(A) = 1$.

$$P(\text{4 of spades}) = P(A) = \frac{1}{52}$$

  (b) Let B be the event 'the card is the 4 of spades or any diamond'. There are 13 diamonds, so $n(B) = 1 + 13 = 14$.

$$P(\text{4 of spades or any diamond}) = P(B) = \frac{14}{52} = \frac{7}{26}$$

(ii) If the 3 of diamonds is placed on the table, there are now 51 cards in the pack, so $n(S) = 51$.

Let R be the event 'the card is from a red suit'. There are now 12 diamonds and 13 hearts in the pack, so $n(R) = 25$.

$$P(\text{from red suit}) = P(R) = \frac{25}{51}$$

Information may be given in a **two-way table**, as in the following example.

Example 3.4

The table shows the results of all the driving tests taken at a particular test centre during the first week of September. A person is chosen at random from those who took their driving test that week.

	Male	Female
Pass	32	43
Fail	10	15

(i) Find the probability that the person passed the driving test.

(ii) Find the probability that the person is a female who failed the driving test.

(iii) A male is chosen. What is the probability that he passed the driving test?

First work out the row and column totals, and the grand total.

	Male	Female	
Pass	32	43	75
Fail	10	15	25
	42	58	**100**

Make sure that the row totals and the column totals give the same 'grand total', shown bold in the table.

(i) P(passed test) $= \dfrac{75}{100} = 0.75$

	Male	Female	
Pass			75
Fail			
			100

(ii) P(female who failed the driving test)

$\qquad = \dfrac{15}{100} = 0.15$

	Male	Female	
Pass			
Fail		15	
			100

(iii) *Now consider only the males.*
P(chosen male passed test)

$\qquad = \dfrac{32}{42} = \dfrac{16}{21}$

	Male	Female	
Pass	32		
Fail			
	42		

Exercise 3a

1 An ordinary fair cubical die is thrown. Find the probability that the score is

　(i) even,

　(ii) lower than 7,

　(iii) a factor of 6,

　(iv) at least 4,

　(v) higher than 1.

2 In a box of 40 highlighters, there are 10 red, 15 blue, 5 green and 10 yellow highlighters. Of these, 8 have dried up and will not write. Ed picks a highlighter at random from the box. Find the probability that the highlighter

　(i) is blue,

　(ii) is neither green nor yellow,

　(iii) is purple,

　(iv) will write.

3 An integer is picked at random from the integers from 1 to 20 inclusive.

A is the event 'the integer is a multiple of 3' and B is the event 'the integer is a multiple of 4'.

Find　(i) P(A)　(ii) P(B').

4 A card is dealt from a well-shuffled ordinary pack of 52 playing cards.

　(i) Find the probability that the card dealt is

　　(a) a Queen,

　　(b) a heart or a diamond,

　　(c) a picture card showing spades.

　(ii) Two cards are dealt and put face-up on the table. They are the 4 of clubs and the 7 of diamonds. A third card is now dealt. What is the probability that it is a club or a 7?

5 Every work day Jamie catches a bus to work. The bus is never early but it is sometimes late.

Jamie decided to record the number of minutes the bus is late over a period of 10 days. Here are his results.

$\qquad 0, 3, 4, 1, 0, 0, 5, 4, 6, 0.$

Find the probability that on a randomly chosen day from the 10 days

　(i) the bus was on time,

　(ii) the bus was more than the mean number of minutes late.

Jamie estimates that the probability that his bus will be late when he is on his way home from work is 0.75.

(iii) What is the probability that his bus will not be late when he is on his way home from work?

6 In a party bag of blue, red, green and yellow balloons, 55 are long balloons and 45 are round balloons.

	Blue	Red	Green	Yellow
Long	12	15	10	a
Round	10	14	b	7

(i) Find the values of a and b.

(ii) Yizi takes a balloon at random from the bag and blows it up. Find the probability that the balloon she takes

(a) is red,

(b) is a blue round balloon,

(c) is not yellow.

7 In a survey, people were asked which factor from the following list influenced them most when buying a car.

A − Price
B − Reliability
C − Fuel economy
D − Servicing costs
E − Range of optional extras

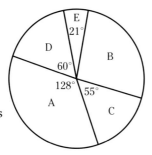

The pie chart shows the results from 90 people.

The names of those who took part were then placed in a prize draw. Find the probability that someone who said 'Reliability' will win the prize.

8 The 30 pupils in a particular class were asked how many brothers and sisters they had. The answers are shown in the table.

Number of brothers and sisters	0	1	2	3	4	5
Number of pupils	4	12	8	3	2	1

Find the probability that a pupil chosen at random from the class comes from a family with three children.

9 A cubical die, numbered 1 to 6, is weighted so that a 6 is twice as likely to occur as any other number.

Find the probability that when the die is thrown the score is

(i) 6,   (ii) odd.

10 When leaving a supermarket, 50 people were asked to take part in a survey on the amount of time they had spent shopping in the supermarket. The results are shown in the following cumulative frequency graph.

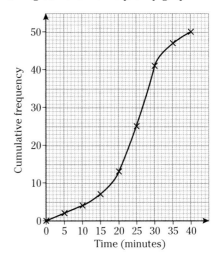

The names of the 50 people were put into a box and one name was picked out at random to receive a prize. Find the probability that the time spent in the supermarket by the prize winner was

(i) less than 20 minutes,

(ii) 35 minutes or more,

(iii) at least 5 minutes but less than 25 minutes,

(iv) more than 45 minutes.

11 An ordinary tetrahedral die has four faces and they are labelled 1, 2, 3, 4. When the die is thrown, the score is the number on which the die lands.
Two fair tetrahedral dice are thrown.
By using a possibility space diagram, or otherwise, find the probability that

(i) the sum of the scores is divisible by 4,

(ii) the product of the scores is an even number,

(iii) the scores differ by at least 2.

12 Two fair coins are tossed together. Find the probability that

    (i) exactly one tail is obtained,

    (ii) at most one head is obtained.

13 Two ordinary fair cubical dice are thrown. Find the probability that

    (i) the sum of the numbers on the dice is 3,

    (ii) the sum of the numbers on the dice exceeds 9,

    (iii) the dice show the same number,

    (iv) the numbers on the dice differ by more than 2.

14 Two ordinary fair cubical dice are thrown at the same time and the scores are multiplied. $P(N)$ denotes the probability that the number N will be obtained.

    (i) Find (a) $P(9)$, (b) $P(4)$, (c) $P(14)$ (d) $P(37)$

    (ii) If $P(N) = \frac{1}{9}$, find the possible values of N.

15 Three fair coins are tossed.

    (i) List all the possible outcomes.

    (ii) Find the probability that two heads and one tail are obtained.

USING ARRANGEMENTS, PERMUTATIONS AND COMBINATIONS

In order to find the number of outcomes in a particular event and in the possibility space, you may need to use arrangements, permutations or combinations.

Example 3.5

Ivan throws three fair dice.

    (i) List all the possible scores on the three dice which give a total of 5, and hence show that the probability of Ivan obtaining a total score of 5 is $\frac{1}{36}$.

    (ii) Find the probability of Ivan obtaining a total score of 7.

                                                 Cambridge Paper 6 Q2 N02

    (i) The possible outcomes giving a score of 5 are

          $(1, 1, 3), (1, 3, 1), (3, 1, 1),$ 3 ways

> Arrangements page **80**
> Number of ways of arranging 3 items of which 2 are alike $= \frac{3!}{2!} = 3$

        and $(1, 2, 2),$ $(2, 1, 2),$ $(2, 2, 1)$ 3 ways

        To give a score of 5, number of outcomes $= 3 + 3 = 6$

        Total number of possible outcomes $= 6 \times 6 \times 6 = 216$

> There are 6 possible outcomes on the first die, 6 on the second and 6 on the third, giving $6 \times 6 \times 6 = 6^3 = 216$.

        $P(\text{total score of 5}) = \frac{6}{216} = \frac{1}{36}$

    (ii) Consider the possible outcomes giving a score of 7:

        $(1, 1, 5), (1, 5, 1), (5, 1, 1)$ 3 ways

        $(1, 3, 3), (3, 1, 3), (3, 3, 1)$ 3 ways

        $(2, 2, 3), (2, 3, 2), (3, 2, 2)$ 3 ways

        $(1, 2, 4), (1, 4, 2), (2, 1, 4), (2, 4, 1), (4, 1, 2), (4, 2, 1)$ 6 ways

> Number of ways of arranging 3 unlike items $= 3! = 6$

        Number of outcomes $= 3 + 3 + 3 + 6 = 15$

        $P(\text{total score of 7}) = \frac{15}{216} = \frac{5}{72}$

Example 3.6

Each of the eleven letters of the word MATHEMATICS is written on a separate card and the cards are laid out in a line.

   (i) Calculate the number of different arrangements of these letters.

   (ii) Find the probability that all the vowels are placed together.

   (i) Total number of different arrangements   Arrangements page **80**

$$= \frac{11!}{2! \times 2! \times 2!} = 4\,989\,600$$

           2 As   2 Ms   2 Ts

   (ii) To find the number of arrangements where the vowels are together, treat the vowels AEAI as one 'item'. Together with the 7 consonants, you now have 8 'items'.

   Number of different arrangements of the 8 'items'

$$= \frac{8!}{2! \times 2!} = 10\,080$$

         2 Ms   2 Ts

   However, the vowels A, E, A, I can be arranged in

$$\frac{4!}{2!} = 12 \text{ ways.}$$

         2 As

   So, number of arrangements with the vowels together

$$= 10\,080 \times 12 = 120\,960$$

$$\text{P(vowels all together)} = \frac{120\,960}{4\,989\,600}$$

$$= 0.02424\ldots = 0.0242 \text{ (3 s.f.)}$$

Example 3.7

Four letters are chosen at random from the letters in the word RANDOMLY. Find the probability that all letters chosen are consonants.

The possibility space consists of all the ways to choose 4 letters from 8 letters without any restrictions,

so $n(S) = \binom{8}{4} = {}_8C_4 = 70.$ On calculator: Combinations page **90**

     $\boxed{8}\,\boxed{{}_nC_r}\,\boxed{4}\,\boxed{=}$

Let E be the event 'all the letters chosen are consonants'. There are 6 consonants R, N, D, M, L, Y, from which to choose 4,

so $n(E) = \binom{6}{4} = {}_6C_4 = 15.$

P(all the letters are consonants) $= \text{P}(E) = \frac{15}{70} = \frac{3}{14}$

Example 3.8

A team of 5 pupils is chosen from a class of 7 girls and 8 boys. Find the probability that the team consists of 3 girls and 2 boys.

The possibility space S consists of all the ways to choose 5 pupils from 15 pupils, with no restrictions,

so $n(S) = \begin{pmatrix} 15 \\ 5 \end{pmatrix} = {}_{15}C_5 = 3003$

Let E be the event 'the team contains 3 girls and 2 boys'.

You need to choose 3 girls from 7 girls and 2 boys from 8 boys.

$n(E) = \begin{pmatrix} 7 \\ 3 \end{pmatrix} \times \begin{pmatrix} 8 \\ 2 \end{pmatrix} = {}_7C_3 \times {}_8C_2 = 980$

P(team consists of 3 girls and 2 boys)

$= P(E) = \dfrac{980}{3003} = \dfrac{140}{429}$

Example 3.9

A staff car park at a school has 13 parking spaces in a row. There are 9 cars to be parked.
  (i)   How many different arrangements are there for parking the 9 cars and leaving 4 empty spaces?
 (ii)   How many different arrangements are there if the 4 empty spaces are next to each other?
(iii)   If the parking is random, find the probability that there will **not** be 4 empty spaces next to each other.

Cambridge Paper 6 Q3 N05

  (i)   Number of arrangements of 9 cars in 13 spaces

$$= {}_{13}P_9 = \frac{13!}{(13 - 9)!} = \frac{13!}{4!} = 259\,459\,200$$

Permutations page **87**

Directly on calculator:

⌷13⌷ ⌷${}_nP_r$⌷ ⌷9⌷ ⌷=⌷

 (ii)   Treat the 4 empty spaces next to each other as one wide space.

There are now 10 'spaces'.

Number of ways to arrange 9 cars in 10 spaces

$$= {}_{10}P_9 = \frac{10!}{(10 - 9)!} = \frac{10!}{1!} = 3\,628\,800$$

(iii)   *First find the probability that the 4 empty spaces are next to each other.*

The number in the possibility space S is the number of different arrangements for parking the 9 cars. From part (i), $n(S) = {}_{13}P_9$.

Let E be the event 'the 4 empty spaces are next to each other'.

From part (ii), $n(E) = {}_{10}P_9$.

P(4 empty spaces are next to each other)

$$= \frac{{}_{10}P_9}{{}_{13}P_9} = 0.01398\ldots$$

So P(4 empty spaces are **not** next to each other)

$$= 1 - 0.01398\ldots$$
$$= 0.9860\ldots$$
$$= 0.986 \text{ (3 s.f.)}$$

$P(A') = 1 - P(A)$

Exercise 3b

1 Each of the 9 letters in the word FACETIOUS is written on a card and the cards are placed in random order in a line.

 (i) How many different arrangements are there?

 (ii) What is the probability that the arrangement begins with F and ends with S?

2 On a shelf there are 4 different mathematics books and 8 different English books.

 (i) The books are to be arranged so that the mathematics books are together. In how many different ways can this be done?

 (ii) What is the probability that all the mathematics books are **not** together?

3 The letters of the word ABSTEMIOUS are arranged in a line at random. Find the probability that the vowels and consonants appear alternately.

4 A bag contains six white counters and eight blue counters. Four counters are chosen at random.

 Find the probability that:

 (i) two white counters and two blue counters are chosen,

 (ii) all the counters are the same colour.

5 A committee consists of 4 women and 2 men. A sub-committee is formed consisting of three members of the committee. Find the probability that the sub-committee consists of:

 (i) 1 man and 2 women,

 (ii) at least 1 man.

6 Two pupils are chosen at random from a class of 10 boys and 8 girls. Find the probability that:

 (i) they are both girls,

 (ii) they are both boys,

 (iii) there is one boy and one girl.

7 The letters of the word PROBABILITY are arranged at random in a line. Find the probability that the two letters I:

 (i) are together,

 (ii) are separated.

8 Four letters are picked at random from the word BREAKDOWN. Find the probability that there is at least one vowel among the letters.

9 Suan is given a bag of 20 sweets of which 6 are apple flavoured, 6 are lemon flavoured and 8 are orange flavoured. Suan takes out 5 sweets at random and eats them. Find the probability that she eats:

 (i) 5 orange flavoured sweets,

 (ii) 3 apple flavoured and 2 lemon flavoured sweets,

 (iii) exactly 2 apple flavoured sweets,

 (iv) no lemon flavoured sweets.

10 Three letters are selected at random from the 9 letters of the word UNIVERSAL. The order in which the letters are selected is unimportant.

 (i) Find the number of selections of 3 letters.

 (ii) Find the probability that the letter V is included in the selection.

11 When a tetrahedral die, with faces labelled 1, 2, 3, 4 is thrown, the score is the number on which it lands.

Jake throws three fair tetrahedral dice.

(i) List all the possible scores on the three dice which give a total of 6.

(ii) Find the probability that Jake obtains a total score of 6.

12 A plate contains 15 cakes of which 6 have yellow icing, 5 have green icing and 4 have pink icing. Three cakes are taken at random from the plate.

Find the probability that:

(i) exactly two of the cakes have green icing,

(ii) one cake has green icing, one has pink icing and one has yellow icing,

(iii) none of the cakes has yellow icing.

13 Jayne has 4 cards.

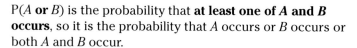

She chooses one or more of these cards and places them on a table to form a number. For example, she could form the number 6 or 31 or 731 or 3671.

(i) Find the total number of different numbers that can be formed.

(ii) If one of the numbers in part (i) is chosen at random, what is the probability that it is greater than 300?

14 Peta deals a hand of 10 cards from a well-shuffled pack of ordinary playing cards. Show that the probability that she deals exactly 5 spades is less than 5%.

TWO OR MORE EVENTS

Now consider two events, A and B, in the possibility space.

P(A **and** B) is the probability that **both A and B** occur.

In set notation P(A and B) is written P($A \cap B$) and read as A intersection B. Although set notation will not be tested in the examination, it can be a concise notation to use.

P(A **and** B) = P($A \cap B$) = P(both A and B occur)

The set of outcomes resulting in both A and B is shown in the Venn diagram by the intersection (overlap) of A and B.

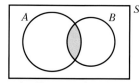

$A \cap B$ means 'A <u>and</u> B'

P(A **or** B) is the probability that **at least one of A and B occurs**, so it is the probability that A occurs or B occurs or both A and B occur.

In set notation P(A or B) is written P($A \cup B$), where $A \cup B$ is read as A union B.

P(A or B) = P($A \cup B$) = P(at least one of A and B occurs)
                            = P(A occurs or B occurs or both occur)

The set of outcomes resulting in A or B, i.e. at least one of A and B, is shown in the Venn diagram by shading the whole of A and the whole of B.

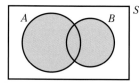

$A \cup B$ means 'A <u>or</u> B <u>or</u> both'

Note that P(A or B) means P(A or B or both). It does **not** mean P(A or B but not both).

Notice that if you add $n(A)$ and $n(B)$ you have counted the intersection (overlap) twice. So, to find the number of outcomes resulting in **A or B**, subtract the number in the intersection, where

$$n(A \text{ or } B) = n(A) + n(B) - n(A \text{ and } B)$$

Now divide each term by $n(S)$:

$$\frac{n(A \text{ or } B)}{n(S)} = \frac{n(A)}{n(S)} + \frac{n(B)}{n(S)} - \frac{n(A \text{ and } B)}{n(S)}$$

This gives the **addition rule for combined events**:

$$P(A \textbf{ or } B) = P(A) + P(B) - P(A \textbf{ and } B)$$

A **or** *B* means
A or *B* or both

A **and** *B* means
both *A* and *B*

Using set notation:

$$P(A \cup B) = P(A) + P(B) + P(A \cap B)$$

Example 3.10

Two events, *X* and *Y*, are such that P(*X* or *Y*) = 0.8, P(*X* and *Y*) = 0.35 and P(*X*) = 0.6.
Find P(*Y'*).

First find P(Y) using the addition rule

$$P(X \text{ or } Y) = P(X) + P(Y) - P(X \text{ and } Y)$$
$$0.8 = 0.6 + P(Y) - 0.35$$
$$P(Y) = 0.55$$

Now find P(Y') using P(Y') = 1 − P(Y)

$$P(Y') = 1 - 0.55$$
$$= 0.45$$

Example 3.11

Some pupils did a survey on comics. They asked all 100 pupils in their year group whether they had read particular comics during the past week. They found that 65 had read Whizz, 55 had read Wham, 30 had read Whizz and Wham and some pupils in the year group had not read either comic.

A pupil was selected at random from the year group to answer more questions in the survey. Find the probability that the pupil

  (i)   had read Whizz or Wham,

  (ii) had not read either comic.

Let *Z* be the event 'the pupil had read Whizz' and *M* be the event 'the pupil had read Wham'.

Method 1: Using the addition rule

  (i)   You know that $P(Z) = \frac{65}{100} = 0.65$, $P(M) = \frac{55}{100} = 0.55$ and $P(Z \text{ and } M) = \frac{30}{100} = 0.3$

You have to find P(Z or M). Remember that this means the probability that the pupil had read at least one of the comics, i.e. Whizz or Wham or both Whizz and Wham.

$$P(Z \text{ or } M) = P(Z) + P(M) - P(Z \text{ and } M)$$
$$= 0.65 + 0.55 - 0.3$$
$$= 0.9$$

In set notation
$$P(Z \cup M) = P(Z) + P(M) - P(Z \cap M)$$

So the probability that the pupil had read Whizz or Wham is 0.9.

  (ii) P(pupil had not read either comic) = 1 − 0.9 = 0.1   Note: P(neither *Z* nor *W*) = 1 − P(*Z* or *W*)

Method 2: Using a Venn diagram

You could use a Venn diagram directly as follows:

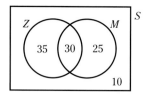

First fill in the 30 who read both by putting them in the intersection (overlap).

You know that 65 read Whizz so fill in the 35 who are in Z but not in the intersection. These only read Whizz.

You know 55 read Wham so fill in the 25 who are in M but not in the intersection. These only read Wham.

From the diagram, you can see that the number in Z or M (i.e. the number who read Whizz or Wham or both) is $35 + 30 + 25 = 90$, so $P(Z \text{ or } M) = \dfrac{90}{100} = 0.9$.

As 90 pupils read one or both comics, the number who had not read either comic is $100 - 90 = 10$, so

$$P(\text{pupil had not read either comic}) = \frac{10}{100} = 0.1$$

Note: It is easy to find other probabilities from the Venn diagram, for example

$$P(\text{pupil had read Whizz but not Wham}) = \frac{35}{100} = 0.35$$

$$P(\text{pupil had read only one of the comics}) = \frac{35 + 25}{100} = 0.6$$

Mutually exclusive events

Events are **mutually exclusive** (or exclusive) if they cannot occur at the same time.

For example, with one throw of a die, the events 'scoring a 3' and 'scoring a 5' are mutually exclusive, since the score cannot be 3 and 5 at the same time. However, the events 'scoring an even number' and 'scoring a prime number' are not mutually exclusive, since a score of 2 is both even and prime.

When A and B are mutually exclusive there is no overlap between A and B,

i.e. $P(A \text{ and } B) = 0$

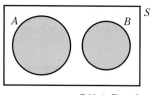

$P(A \cap B) = 0$

So, the addition rule for mutually exclusive events is

$$P(A \textbf{ or } B) = P(A) + P(B)$$

This is sometimes known as the '**or**' rule for mutually exclusive events.

In set notation $P(A \cup B) = P(A) + P(B)$

The rule can be extended to n mutually exclusive events $A_1, A_2, A_3, \ldots, A_n$ as follows:

$$P(A_1 \text{ or } A_2 \text{ or } A_3 \text{ or } \ldots \text{ or } A_n) = P(A_1) + P(A_2) + P(A_3) + \ldots + P(A_n).$$

Example 3.12

In a race where there can be only one winner, the probability that John will win is 0.3, the probability that Paul will win is 0.2 and the probability that Mark will win is 0.4.

Find the probability that

(i)   John or Mark wins,

(ii) John or Paul or Mark wins,

(iii) someone else wins.

Only one person wins, so the events are mutually exclusive.

(i)   P(John or Mark wins) = 0.3 + 0.4 = 0.7

(ii) P(John or Mark or Paul wins) = 0.3 + 0.4 + 0.2 = 0.9

(iii) P(someone else wins) = 1 − 0.9 = 0.1

Example 3.13

A card is dealt from an ordinary pack of 52 playing cards. Find the probability that the card is

(i) a club or a diamond,

(ii) a club or a King.

Possibility space S is the pack of 52 cards, so $n(S) = 52$

(i) C is the event 'the card is a club'.

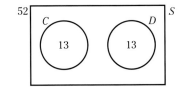

$$P(C) = \frac{n(C)}{n(S)} = \frac{13}{52} = \frac{1}{4}$$

D is the event 'the card is a diamond'.

$$P(D) = \frac{n(D)}{n(S)} = \frac{13}{52} = \frac{1}{4}$$

Since a card cannot be both a club and a diamond, the events C and D are **mutually exclusive**.

So $P(C \text{ or } D) = P(C) + P(D)$

$$= \frac{1}{4} + \frac{1}{4} = \frac{1}{2}$$

(ii) K is the event 'the card is a King'.

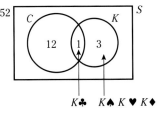

$$P(K) = \frac{4}{52} = \frac{1}{13}$$

The events C and K are **not mutually exclusive** since a card can be both a King and a club,

Now $P(C \text{ and } K) = P(\text{King of clubs}) = \frac{1}{52}$

Therefore

$$P(C \text{ or } K) = P(C) + P(K) - P(C \text{ and } K)$$

$$= \frac{13}{52} + \frac{4}{52} - \frac{1}{52}$$

$$= \frac{16}{52} = \frac{4}{13}$$

If you are asked to **show** that events A and B are mutually exclusive, you must give working to show that either of the following is satisfied:

$$P(A \text{ or } B) = P(A) + P(B)$$ $$P(A \text{ and } B) = 0$$

Example 3.14

Two fair dice are thrown.

(i) Event A is 'the scores differ by 3 or more'. Find the probability of event A.

(ii) Event B is 'the product of the scores is greater than 8'. Find the probability of event B.

(iii) State with a reason whether events A and B are mutually exclusive.

Cambridge Paper 6 Q4 N06

Show the outcomes on a possibility space diagram.

There are 36 possible outcomes in the possibility space S.

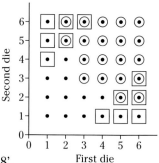

(i) Let A be the event 'the scores differ by 3 or more'.

There are 12 outcomes where the scores differ by 3 or more. These are shown in the possibility space with a square.

So $P(A) = \dfrac{12}{36} = \dfrac{1}{3}$

(ii) Let B be the event 'the product of the scores is greater than 8'.

There are 20 outcomes where the product is greater than 8. These are shown circled in the possibility space.

So $P(B) = \dfrac{20}{36} = \dfrac{5}{9}$

(iii) There are 6 outcomes resulting in both A and B, shown with a square and a circle,

so $P(A \text{ and } B) = \dfrac{6}{36} = \dfrac{1}{6}$

Since $P(A \text{ and } B) \neq 0$, A and B are not mutually exclusive.

Alternatively, you could show that $P(A \text{ or } B) \neq P(A) + P(B)$.

Exercise 3c

*Many of the questions in this exercise can be done directly by using a possibility space, Venn diagram or two-way table. Take care, however, to distinguish between P(A **or** B), P(A **and** B).*

1 An ordinary fair cubical die is thrown. Find the probability that the score on the die is

 (i) even,

 (ii) prime,

 (iii) even or prime,

 (iv) even and prime.

2 All the students in a class of 30 study at least one of the subjects, physics and biology. Of these, 20 study physics and 21 study biology. Find the probability that a student chosen at random (i) studies both physics and biology, (ii) studies only physics.

3 From an ordinary pack of 52 playing cards the seven of diamonds has been lost. A card is dealt from the well-shuffled pack. Find the probability that the card is

 (i) a diamond,

 (ii) a Queen,

 (iii) a diamond or a Queen,

 (iv) a diamond or a seven,

 (v) a diamond and a seven.

4 In a quality control test, all the components produced by three machines on a particular day are tested. The results are summarised in the two-way table.

	Machine A	Machine B	Machine C
Faulty	2	3	1
Not faulty	80	72	42

A machine operator picks up one of these components at random. Find the probability that the component

(i) is from Machine A,

(ii) is a faulty component from Machine C,

(iii) is not faulty or is from Machine A.

5 A class consists of 9 boys and 11 girls. Of these, 4 boys and 3 girls are in the athletics team. A pupil is chosen at random from the class to take part in the 'egg and spoon' race on Sports Day. Find the probability that the pupil is

(i) in the athletics team,

(ii) a girl,

(iii) a girl in the athletics team,

(iv) a girl or in the athletics team.

6 Events C and D are such that $P(C) = \frac{19}{30}$, $P(D) = \frac{2}{5}$ and $P(C \text{ or } D) = \frac{4}{5}$.
Find $P(C \text{ and } D)$.

7 Events X and Y are such that $P(X) = 0.75$, $P(Y') = 0.45$ and $P(X \text{ and } Y) = 0.5$.
Find $P(X \text{ or } Y)$.

8 Events A and B are such that $P(A) = P(B)$, $P(A \text{ and } B) = 0.1$ and $P(A \text{ or } B) = 0.7$. Find $P(A')$.

9 Events A and B are such that
$P(A \text{ occurs}) = 0.6$, $P(B \text{ occurs}) = 0.7$,
$P(\text{at least one of } A \text{ and } B \text{ occurs}) = 0.9$.

Find

(i) P(both A and B occur),

(ii) P(neither A nor B occurs),

(iii) P(A occurs or B occurs but not both A and B occur).

10 X and Y are mutually exclusive events such that $P(X) = 0.5$ and $P(Y) = 0.25$.
Find   (i) P(X or Y),   (ii) P(X and Y).

11 The probability that a randomly chosen boy in Class 2 is in the football team is 0.4, the probability that he is in the chess team is 0.5 and the probability that he is in both teams is 0.2.

Find the probability that a boy chosen at random from the class

(i) is in the football team, but not in the chess team,

(ii) is in the football team or the chess team,

(iii) is not in either team.

12 Two fair cubical dice are thrown. Find the probability that the sum of the scores is

(i) a multiple of 5,

(ii) greater than 9,

(iii) a multiple of 5 or greater than 9,

(iv) a multiple of 5 and greater than 9.

13 A fair cubical die is thrown. Events are defined as follows:
A: the score is at most 3
B: the score is at least 3
C: the score is lower than 3
D: the score is higher than 3

(i) Identify pairs of events that are mutually exclusive.

(ii) Find P(A or B).

(iii) Find P(A and C).

14 Two fair coins are tossed.

(i) Events A and B are mutually exclusive. A is the event 'at least one head is obtained'. Define event B.

(ii) X is the event 'one head is obtained'. Define an event Y such that X and Y are not mutually exclusive.

15 Two fair cubical dice are thrown.
Event A is 'the scores on the dice are the same'.
Event B is 'the product of the scores is a multiple of 3'.
Event C is 'the sum of the scores is 7'.
State, with a reason, whether the following pairs of events are mutually exclusive:

(i) A and B

(ii) A and C

(iii) B and C.

CONDITIONAL PROBABILITY

Conditional probability is used when the probability that an event will occur depends on whether another event has occurred.

For events A and B, the **conditional probability** that event B occurs, given that event A has already occurred, i.e. P(B given A) is written $P(B \mid A)$.

Since A has already occurred, the possibility space is reduced to just A.

So, $P(B \mid A) = \dfrac{n(A \text{ and } B)}{n(A)}$

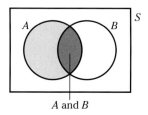

Now divide the top and bottom of the fraction by $n(S)$:

$$P(B \mid A) = \dfrac{\dfrac{n(A \text{ and } B)}{n(S)}}{\dfrac{n(A)}{n(S)}}$$

$$= \dfrac{P(A \text{ and } B)}{P(A)}$$

This gives the **multiplication** rule:

$$P(A \text{ and } B) = P(A) \times P(B \mid A)$$

Example 3.15

There are 5 red counters and 7 blue counters in a bag. Darian takes a counter from the bag, puts it on the table and then takes another counter from the bag. Find the probability that he takes out

(i)   a red counter then a blue counter,

(ii) two counters that are the same colour,

(iii) a least one red counter.

5 R, 7 B

(i) $P(R_1 \text{ and } B_2) = P(R_1) \times P(B_2 \mid R_1)$

$$= \frac{5}{12} \times \frac{7}{11}$$

$$= \frac{35}{132}$$

> This is known as **sampling without replacement** as the first counter is not put back into the bag before the second counter is taken out.

> There are now only 11 counters in the bag, 7 of which are blue

(ii) $P(\text{same colour}) = P(\text{both red}) + P(\text{both blue})$

$$= P(R_1) \times P(R_2 \mid R_1) + P(B_1) \times P(B_2 \mid B_1)$$

$$= \frac{5}{12} \times \frac{4}{11} + \frac{7}{12} \times \frac{6}{11}$$

$$= \frac{5}{33} + \frac{7}{22}$$

$$= \frac{31}{66}$$

(iii) $P(\text{at least one red}) = 1 - P(\text{both red})$

$$= 1 - \frac{5}{33} \quad \text{from part (ii)}$$

$$= \frac{28}{33}$$

> You could use
>
> P(at least one red)
>
> $= P(\text{both red}) + P(R_1 \text{ and } B_2) + P(B_1 \text{ and } R_2)$

Example 3.16

Two events X and Y are such that $P(X) = 0.2$, $P(Y) = 0.25$, $P(Y|X) = 0.4$.

Find   (i) $P(X \text{ and } Y)$,   (ii) $P(X|Y)$,   (iii) $P(X \text{ or } Y)$.

(i) $P(X \text{ and } Y) = P(X) \times P(Y|X)$
$$= 0.25 \times 0.4$$
$$= 0.1$$

(ii) $P(X|Y) = \dfrac{P(X \text{ and } Y)}{P(Y)}$
$$= \dfrac{0.1}{0.2}$$
$$= 0.5$$

(iii) $P(X \text{ or } Y) = P(X) + P(Y) - P(X \text{ and } Y)$
$$= 0.2 + 0.25 - 0.1$$
$$= 0.35$$

Example 3.17

Of the 120 first year students at a college, 36 study chemistry, 60 study biology and 10 study both chemistry and biology. A first year student is selected at random to represent the college at a conference. Find the probability that the student studies

(i) chemistry, given that the student studies biology,
(ii) biology, given that the student studies chemistry.

Event C: student studies chemistry
Event B: student studies biology

You are given that

$$P(C) = \frac{36}{120}, P(B) = \frac{60}{120}, P(C \text{ and } B) = \frac{10}{120}$$

(i) $P(C|B) = \dfrac{P(C \text{ and } B)}{P(B)} = \dfrac{\frac{10}{120}}{\frac{60}{120}} = \dfrac{1}{6}$

(iii) $P(B|C) = \dfrac{P(B \text{ and } C)}{P(C)} = \dfrac{\frac{10}{120}}{\frac{36}{120}} = \dfrac{10}{36} = \dfrac{5}{18}$

Note that you could do this question directly, using a Venn diagram:

Fill in the 10 who study both by putting them in the intersection (overlap), then work out the number just in C (and not in the intersection) and the number just in B (and not in the intersection)

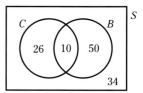

(i) To find $P(C \mid B)$, look only at the reduced possibility space of B.

$$P(C \mid B) = \frac{n(C \text{ and } B)}{n(B)} = \frac{10}{60} = \frac{1}{6}$$

(ii) To find $P(B \mid C)$ look only at the reduced possibility space of C.

$$P(B \mid C) = \frac{n(B \text{ and } C)}{n(C)} = \frac{10}{36} = \frac{5}{18}$$

Example 3.18

Last month a consultant saw 60 men and 65 women suspected of having a particular eye condition. Tests were carried out and the following table shows the results. The totals are shown in bold.

	Had eye condition (C)	Did not have eye condition (C')	
Man (M)	25	35	**60**
Woman (W)	20	45	**65**
	45	**80**	**125**

One of these patients was selected at random to take part in a survey. Find the probability that the patient selected

(i) was a woman, given that the patient had the eye condition,

(ii) had the eye condition, given that the patient was a man.

Focus on the relevant numbers in the table.

(i) P(woman, given eye condition)

$= P(W \mid C)$

$= \dfrac{20}{45} = \dfrac{4}{9}$

	C	C'	
M			
W	20		
	45		

The possibility space has been reduced to the 45 patients in C.

(ii) P(did not have eye condition, given man)

$= P(C' \mid M)$

$= \dfrac{35}{60} = \dfrac{7}{12}$

	C	C'	
M		35	60
W			

The possibility space has been reduced to the 60 men.

Independent events

You know that, in general, for events A and B

$P(A \text{ and } B) = P(A) \times P(B \mid A)$.

Now when either of these events can occur without being affected by the outcome of the other, the events are said to be independent.

So, for **independent events** A and B,

$P(B \mid A) = P(B)$

Also $P(A \mid B) = P(A)$

This gives the **multiplication rule for independent events**:

$P(A \textbf{ and } B) = P(A) \times P(B)$

In set notation $P(A \cap B) = P(A) \times P(B)$

This is sometimes known as the '**and**' rule for independent events.

The rule can be extended to n independent events $A_1, A_2, A_3, \ldots, A_n$ as follows:

$P(A_1 \text{ and } A_2 \text{ and } A_3 \text{ and } \ldots \text{ and } A_n) = P(A_1) \times P(A_2) \times P(A_3) \times \ldots \times P(A_n)$.

Example 3.19

There are 5 red counters and 7 blue counters in a bag. Eliza takes a counter from the bag, notes its colour and then puts it back into the bag. Freddie then takes a counter from the bag. Find the probability that

(i)   Eliza takes a red counter and Freddie takes a blue counter,

(ii) Freddie's counter is the same colour as Eliza's counter.

As Eliza's counter is put back into the bag, the events are independent so, for both Eliza and Freddie,

This is known as **sampling with replacement** as the first counter is put back into the bag before the second counter is taken out.

$P(R) = \dfrac{5}{12}, P(B) = \dfrac{7}{12}$

(i)   P(Eliza picks red and Freddie picks blue)

$= P(R_E) \times P(B_F)$

$= \dfrac{5}{12} \times \dfrac{7}{12}$

$= \dfrac{35}{144}$

There are still 12 counters in the bag, 7 of which are blue

(ii) P(same colour) = P(both red) + P(both blue)

$= P(R_E) \times P(R_F) + P(B_E) \times P(B_F)$

$= \dfrac{5}{12} \times \dfrac{5}{12} + \dfrac{7}{12} \times \dfrac{7}{12}$

$= \dfrac{37}{72}$

Example 3.20

Two fair cubical dice are thrown. One is red and the other is blue. Find the probability that

(i) the score is 3 on both dice,

(ii) neither die has a score of 3.

The scores are independent.

For both the red die and the blue die: $P(3) = \frac{1}{6}$, $P(\text{not } 3) = \frac{5}{6}$

(i) $P(3 \text{ on both dice}) = \frac{1}{6} \times \frac{1}{6} = \frac{1}{36}$

(ii) $P(\text{neither score is } 3) = \frac{5}{6} \times \frac{5}{6} = \frac{25}{36}$

Example 3.21

The probability that a certain type of machine will break down in the first month of operation is 0.1. Three machines of this type are installed at the same time. The performances of the three machines are independent. Find the probability that at the end of the first month

(i) all three machines have broken down,

(ii) just one machine has broken down,

(iii) at least one machine is working.

Let B_n be the event 'the nth machine has broken down'
and W_n be the event 'the nth machine is working'.

(i) P(all three machines have broken down)

$$= P(B_1 B_2 B_3)$$
$$= 0.1 \times 0.1 \times 0.1$$
$$= 0.001$$

(ii) P(just one machine has broken down)

$$= P(B_1 W_2 W_3) + P(W_1 B_2 W_3) + P(W_1 W_2 B_3)$$
$$= (0.1 \times 0.9 \times 0.9) + (0.9 \times 0.1 \times 0.9) + (0.9 \times 0.9 \times 0.1)$$
$$= 0.243$$

(iii) P(at least one machine is working)

$$= 1 - P(\text{all three have broken down})$$
$$= 1 - 0.001 \qquad \text{from part (i)}$$
$$= 0.999$$

Although it would be more time consuming, you could work out the following:
P(at least one is working)
= P(just one is working) + P(just two are working) + P(all three are working)

If you are asked to **show** that events A and B are **independent** you must give working to show that any one of the following is satisfied:

$P(A \text{ and } B) = P(A) \times P(B)$	$P(A \mid B) = P(A)$	$P(B \mid A) = P(B)$

Example 3.22

Data about employment for males and females in a small rural area are shown in the table.

	Unemployed	Employed
Male	206	412
Female	358	305

A person from this area is chosen at random. Let M be the event that the person is male and let E be the event that the person is employed.

(i) Find P(M).

(ii) Find P(M and E).

(iii) Are M and E independent events? Justify your answer.

(iv) Given that the person chosen is unemployed, find the probability that the person is female.

Cambridge Paper 6 Q5 J05

First find the row and column totals and the grand total.

Remember to check that the row totals and the column totals give the same grand total.

	Unemployed (U)	Employed (E)	
Male (M)	206	412	**618**
Female (F)	358	305	**663**
	564	**717**	**1281**

Now use the totals to work out the probabilities.

Let M be the event 'male', F be the event 'female', U be the event 'unemployed' and E be the event 'employed'.

(i) $\mathrm{P}(M) = \dfrac{618}{1281} = 0.4824\ldots = 0.482$ (3 s.f.)

	U	E	
M			618
F			
			1281

(ii) $\mathrm{P}(M \text{ and } E) = \dfrac{412}{1281} = 0.3216\ldots = 0.322$ (3 s.f.)

	U	E	
M		412	
F			
			1281

(iii) If events M and E are independent, then P(M and E) = P(M) $\times$ P(E).

$\mathrm{P}(E) = \dfrac{717}{1281}$

	U	E	
M			
F			
		717	1281

Now $\mathrm{P}(M) \times \mathrm{P}(E) = \dfrac{618}{1281} \times \dfrac{717}{1281} = 0.2700\ldots$

whereas $\mathrm{P}(M \text{ and } E) = 0.3216\ldots$ from part (ii)

Since P(M and E) $\neq$ P(M) $\times$ P(E), M and E are **not** independent events.

Alternatively

$P(M \mid E) = \frac{412}{717} = 0.5746\ldots$

$P(M) = 0.4824\ldots$

	U	E	
M		412	
F			
		717	

Since $P(M \mid E) \neq P(M)$, M and E are not independent events.

Showing that $P(E \mid M) \neq P(E)$ also shows that M and E are not independent events.

(iv) $P(F \mid U) = \frac{358}{564} = 0.635$ (3 s.f.)

	U	E	
M			
F	358		
	564		

Example 3.23

Events A and B are such that $P(A) = 0.3$, $P(B) = 0.6$ and $P(A \text{ or } B) = 0.72$. State, giving a reason in each case, whether events A and B are

(i) mutually exclusive,

(ii) independent.

$P(A) = 0.3$, $P(B) = 0.6$

(i) If A and B are mutually exclusive then

$$P(A) + P(B) = P(A \text{ or } B)$$

$P(A) + P(B) = 0.3 + 0.6 = 0.9$

$P(A \text{ or } B) = 0.72$

So $P(A) + P(B) \neq P(A \text{ or } B)$ and events A and B are not mutually exclusive.

(ii) If events A and B are independent, $P(A \text{ and } B) = P(A) \times P(B)$

First find $P(A \text{ and } B)$.

$$P(A \text{ or } B) = P(A) + P(B) - P(A \text{ and } B)$$

i.e. $0.72 = 0.3 + 0.6 - P(A \text{ and } B)$

so $P(A \text{ and } B) = 0.18$

Now check $P(A) \times P(B)$

$P(A) \times P(B) = 0.3 \times 0.6 = 0.18$

Since $P(A \text{ and } B) = P(A) \times P(B)$, events A and B are independent.

Exercise 3d

1 Each of the numbers 1, 2, 3, 4, 5, 6, 7, 8, 9 is written on a card and the nine cards are shuffled. A card is then dealt.

Given that the card is a multiple of 3, find the probability that the card is

(i) even,

(ii) a multiple of 4.

2 A bag contains 20 identical sweets apart from the colour: 10 are pink, 7 are green and 3 are yellow. Jovian randomly selects two sweets from the bag, one after the other, and eats them. Find the probability that

(i) she eats two pink sweets,

(ii) the first sweet is green and the second sweet is yellow,

(iii) she eats exactly one pink sweet,

(iv) neither sweet is green.

3 A fair cubical die is thrown twice. Find the probability of obtaining

(i) a score of 2 on both throws,

(ii) a score of 2 on just one throw,

(iii) a score of 4 on at least one throw,

(iv) a score lower than 3 on both throws.

4 In a large group of people it is known that 10% have a hot breakfast, 20% have a hot lunch and 25% have a hot breakfast or a hot lunch. Find the probability that a person chosen at random from this group

(i) has a hot breakfast and a hot lunch,

(ii) has a hot lunch, given that the person chosen had a hot breakfast.

5 A card is picked from an ordinary pack containing 52 playing cards. It is then replaced in the pack, the pack is shuffled and a second card is picked. Find the probability that

(i) both cards are the seven of diamonds,

(ii) the first card is a heart and the second card is a spade,

(iii) at least one card is a Queen.

6 Two fair tetrahedral dice, each with faces labelled 1, 2, 3, 4, are thrown and the number on which each lands is noted.

(i) Find the probability that the sum of the two numbers is even, given that at least one die lands on a 3.

(ii) Find the probability that at least one die lands on a 3, given that the score is even.

7   In a group of 100 college students, 80 own a laptop computer, 65 own a desktop computer and 50 own both a laptop computer and a desktop computer. Find the probability that a student chosen at random from the group

   (i) owns a desktop computer, given that the student owns a laptop computer,

   (ii) does not own a laptop computer, given the student owns a desktop computer,

   (iii) does not own a laptop computer or a desktop computer.

8   The events A and B are such that $P(A \mid B) = 0.4$, $P(B \mid A) = 0.25$ and $P(A$ and $B) = 0.12$.

   (i) Are A and B independent? Give a reason for your answer.

   (ii) Find $P(A$ or $B)$.

9   Events A and B are such that $P(A) = 0.45$, $P(B) = 0.35$ and $P(A$ or $B) = 0.7$.

   (i) Find $P(A$ and $B)$.

   (ii) Show that events A and B are not independent.

   (iii) Find $P(A \mid B)$.

10 Two ordinary fair dice, one red and one blue, are thrown.

   Events A, B and C are defined as follows:

   Event A: the number showing on the red die is 5 or 6

   Event B: the total of the numbers showing on the two dice is 7

   Event C: the total of the numbers showing on the two dice is 8.

   (i) State, with a reason, which two of the events A, B and C are mutually exclusive.

   (ii) Show that the events A and B are independent.

11 A school has 100 teachers. In a survey on the use of the school car park, the teachers were asked whether they had driven a car to school on a particular day. Of the 70 full-time teachers, 45 had driven a car to school and of the 30 part-time teachers, 12 had driven a car to school.

   (i) Copy and complete the two-way table, where C denotes the event 'the teacher had driven a car to school that day'.

	C	C'	Total
Full-time teacher			
Part-time teacher			
Total			100

   (ii) Find the probability that a teacher chosen at random

   (a) is a part-time teacher who had driven a car to school,

   (b) is a full-time teacher who had not driven a car to school,

   (c) is a full-time teacher or had driven a car to school,

   (d) is a part-time teacher, given that the teacher had driven a car to school.

   (iii) Are the events 'the teacher had driven a car to school' and 'the teacher is full-time' independent? Give a reason for your answer.

   (iv) Describe two events that are mutually exclusive.

12 Each father in a random sample of fathers was asked how old he was when his first child was born. The following histogram represents the information.

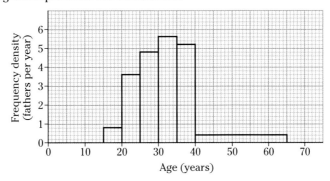

(i) What was the modal group?

(ii) How many fathers were between 25 and 30 years old when their first child was born?

(iii) How many fathers were in the sample?

(iv) Find the probability that a father, chosen at random from the group, was between 25 and 30 years old when his first child was born, given that he was older than 25 years.

Cambridge Paper 6 Q5 J06

PROBABILITY TREES

A very useful method for tackling many probability problems is to draw a **tree diagram**. It can be used when events are conditional and when they are independent. The method is described in the following example.

Example 3.24

In country A 30% of people who drink tea have sugar in it. In country B 65% of people who drink tea have sugar in it. There are 3 million people in country A who drink tea and 12 million people in country B who drink tea. A person is chosen at random from these 15 million people.

(i) Find the probability that the person chosen is from country A.

(ii) Find the probability that the person chosen does not have sugar in their tea.

(iii) Given that the person chosen does not have sugar in their tea, find the probability that the person is from country B.

Cambridge Paper 6 Q2 J08

(i) Let A be the event 'the person is from country A'.

$$P(A) = \frac{3 \text{ million}}{15 \text{ million}} = \frac{3}{15} = 0.2$$

(ii) *To draw a tree diagram, first draw a set of branches to show the events 'the person is from country A' and 'the person is from country B'. Write in the probabilities that you know:*

$$P(A) = 0.2, P(B) = 1 - 0.2 = 0.8$$

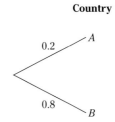

Now draw a set of branches from A to show the events 'person from A has sugar' and 'person from A does not have sugar'.

Let S be the event 'the person has sugar in tea'.

You know that $P(S \mid A) = 30\% = 0.3$

So $P(S' \mid A) = 1 - 0.3 = 0.7$

This is an example of **conditional events**.

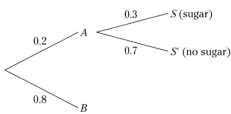

Note that the probabilities on the branches meeting at a point must add up to 1.

Now complete the tree by drawing a set of branches from B to show the events 'person from B has sugar' and 'person from B does not take sugar'.

You know that $P(S \mid B) = 65\% = 0.65$

So $P(S' \mid B) = 1 - 0.65 = 0.35$

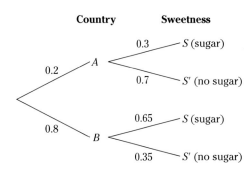

*You can calculate the probabilities of the end results by **multiplying** the probabilities along the branches.*

$P(A \text{ and } S) = P(A) \times P(S \mid A) = 0.2 \times 0.3 = 0.06$

$P(A \text{ and } S') = P(A) \times P(S' \mid A) = 0.2 \times 0.7 = 0.14$

$P(B \text{ and } S) = P(B) \times P(S \mid B) = 0.8 \times 0.65 = 0.52$

$P(B \text{ and } S') = P(B) \times P(S' \mid B) = 0.8 \times 0.35 = 0.28$ *

P(person chosen does not have sugar in tea)

Notice that $0.06 + 0.14 + 0.52 + 0.28 = 1$
The sum of the probabilities of the end outcomes is always 1.

$= P(S')$

$= P(A \text{ and } S') + P(B \text{ and } S')$

Add the probabilities of the outcomes shown shaded.

$= 0.2 \times 0.7 + 0.8 \times 0.35$

$= 0.14 + 0.28$

$= 0.42$

(iii) P(from country B | does not have sugar in tea)

$= P(B \mid S')$ Marked * on the tree

$= \dfrac{P(B \text{ and } S')}{P(S')}$

The **possibility space** has been reduced to the shaded outcomes.

$= \dfrac{0.28}{0.42}$ From part (ii), shaded on tree

$= \dfrac{2}{3}$

You could approximate to give the answer 0.667 (3 s.f.), but $\frac{2}{3}$ is more accurate as it is exact.

Example 3.25

There are three sets of traffic lights on Karinne's journey to work. The independent probabilities that Karinne has to stop at the first, second and third sets of lights are 0.4, 0.8 and 0.3 respectively.

(i) Draw a tree diagram to show this information.

(ii) Find the probability that Karinne has to stop at each of the first two sets of lights but does not have to stop at the third set of lights.

(iii) Find the probability that Karinne has to stop at exactly two of the three sets of lights.

(iv) Find the probability that Karinne has to stop at the first set of lights, given that she has to stop at exactly two sets of lights.

Cambridge Paper 6 Q6 N08

(i) *The first set of branches shows the outcome at the first set of lights, the second set of branches shows the outcomes at the second set of lights and the third set of branches shows the outcomes at the third set of lights.*

*Note that, since the events are **independent**, Karinne stopping at a set of lights is not affected by whether or not she stops at other sets of lights.*

This is an example of **independent events**.

Let S_n represent the event 'Stop at nth set of lights' and G_n represent the event 'Go at nth set of lights'

part (ii)

$P(S_1 S_2 G_3) = 0.4 \times 0.8 \times 0.7 = 0.224$ *

$P(S_1 G_2 S_3) = 0.4 \times 0.2 \times 0.3 = 0.024$ *

$P(G_1 S_2 S_3) = 0.6 \times 0.8 \times 0.3 = 0.144$

(ii) P(Karinne has to stop at each of the first two lights but not at the third)

$$= P(S_1 S_2 G_3)$$

$$= 0.4 \times 0.8 \times 0.7$$

$$= 0.224$$

(iii) P(Karinne has to stop at exactly two sets)

$\qquad = P(S_1 S_2 G_3) + P(S_1 G_2 S_3) + P(G_1 S_2 S_3)$ ←——Shaded on tree

$\qquad = 0.224 + 0.4 \times 0.2 \times 0.3 + 0.6 \times 0.8 \times 0.3$

$\qquad = 0.224 + 0.024 + 0.144$

$\qquad = 0.392$

(iv) P(stop at first set | stops at exactly two sets)

$$= \frac{P(S_1 S_2 G_3) + P(S_1 G_2 S_3)}{P(\text{stops at exactly two sets})}$$ ←——Marked * on tree.
←——From part (iii)

$$= \frac{0.224 + 0.024}{0.392}$$

$$= 0.6326...$$

$$= 0.633 \ (3 \ \text{s.f.})$$

Example 3.26

Dan is playing a game in which players pick counters at random, one at a time without replacement, from a bag. At the beginning of the game, the bag contains 6 red counters and 4 blue counters. Dan takes two counters from the bag.

 (i) Find the probability that both counters are blue.

 (ii) Find the probability that the counters are the same colour.

 (iii) Given that the counters are the same colour, find the probability that they are both blue.

In the tree diagram, first draw a set of branches showing the outcomes for the colour of the first counter, then draw a second set of branches from each of these, showing the outcome for the colour of the second counter.

The probabilities for the second counter depend on the first counter drawn. For example, if the first counter drawn is red then $P(R_1) = \frac{6}{10}$ and, since there are only 9 counters left of which 5 are red, $P(R_2 \mid R_1) = \frac{5}{9}$

This is an example of **conditional events**.

(i)

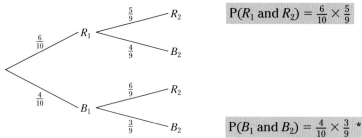

$P(R_1 \text{ and } R_2) = \frac{6}{10} \times \frac{5}{9}$

$P(B_1 \text{ and } B_2) = \frac{4}{10} \times \frac{3}{9}$ *

P(both counters are blue) = P(B_1 and B_2)

$$= \frac{4}{10} \times \frac{3}{9} \qquad \text{Marked * on diagram}$$

$$= \frac{2}{15}$$

(ii) P(same colour) = P(both red) + P(both blue)

$$= \frac{6}{10} \times \frac{5}{9} + \frac{2}{15} \qquad \text{Shaded on diagram}$$

$$= \frac{7}{15}$$

(iii) P(both blue | same colour)

$$= \frac{\text{P(both blue)}}{\text{P(same colour)}}$$

$$= \frac{\frac{2}{15}}{\frac{7}{15}}$$

$$= \frac{2}{7}$$

Example 3.27

When a farmer's dog is let loose, it chases either ducks with probability $\frac{3}{5}$ or geese with probability $\frac{2}{5}$. If the dog chases the ducks, there is a probability of $\frac{1}{10}$ that they will attack the dog. If the dog chases the geese, there is a probability of $\frac{3}{4}$ that they will attack the dog. Given that the dog is not attacked, find the probability that it was chasing the geese.

Cambridge Paper 62 Q6 part (ii) J10

Draw a tree diagram so that the first set of branches shows the type of bird chased: ducks or geese, and the second set of branches shows whether the dog is attacked or not.

$P(D) = \frac{3}{5}$, $P(A \mid D) = \frac{1}{10}$, $P(A' \mid D) = 1 - \frac{1}{10} = \frac{9}{10}$

$P(G) = \frac{2}{5}$, $P(A \mid G) = \frac{3}{4}$, $P(A' \mid G) = 1 - \frac{3}{4} = \frac{1}{4}$

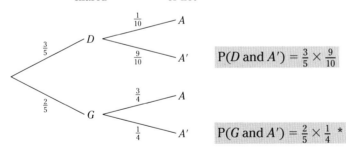

P(dog is chasing geese | dog is not attacked) = $P(G \mid A')$.

Now $P(G \mid A') = \dfrac{P(G \text{ and } A')}{P(A')}$

First find P(G and A')

$P(G \text{ and } A') = \frac{2}{5} \times \frac{1}{4} = \frac{1}{10}$ Marked * on diagram

Now find P(A')

$P(A') = P(D \text{ and } A') + P(G \text{ and } A')$ Shaded on diagram

$\quad = \frac{3}{5} \times \frac{9}{10} + \frac{2}{5} \times \frac{1}{4}$

$\quad = \frac{16}{25}$

So $P(G \mid A') = \dfrac{\frac{1}{10}}{\frac{16}{25}} = \frac{5}{32}$

Example 3.28

The probability that Henk goes swimming on any day is 0.2. On a day that he goes swimming, the probability that Henk has burgers for supper is 0.75. On a day when he does not go swimming, the probability that he has burgers for his supper is x. The information is shown on the following tree diagram.

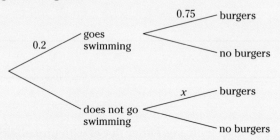

The probability that Henk has burgers for supper on any day is 0.5.

(i) Find x.

(ii) Given that Henk has burgers for supper, find the probability that he went swimming that day.

Cambridge Paper 6 Q2 J06

Let S be the event 'Henk goes swimming' and B be the event 'Henk has burgers'.

(i) You are told that $P(B) = 0.5$

But, from the tree diagram,

$P(B) = P(S \text{ and } B) + P(S' \text{ and } B)$
$= 0.2 \times 0.75 + 0.8 \times x$
$= 0.15 + 0.8x$

So $0.15 + 0.8x = 0.5$
$0.8x = 0.35$
$x = \dfrac{0.35}{0.8} = 0.4375$

$P(S \text{ and } B) = 0.2 \times 0.75$ *

$P(S' \text{ and } B) = 0.8 \times x$

(ii) $P(S \mid B) = \dfrac{P(S \text{ and } B)}{P(B)}$

$= \dfrac{0.2 \times 0.75}{0.5}$

$= 0.3$

Example 3.29

When Don plays tennis, 65% of his first serves go into the correct area of the court. If the first serve goes into the correct area, his chance of winning the point is 90%. If his first serve does not go into the correct area, Don is allowed a second serve and, of these, 80% go into the correct area. If the second serve goes into the correct area, his chance of winning the point is 60%. If neither serve goes into the correct area, Don loses the point.

(i) Draw a tree diagram to represent this information.

(ii) Using your tree diagram, find the probability that Don loses the point.

(iii) Find the conditional probability that Don's first serve went into the correct area, given that he loses the point.

Cambridge Paper 62 Q6(ii) J10

(i) Tree diagram:

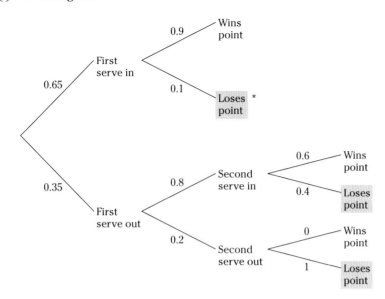

(ii) P(Don loses point)

$$= (0.65 \times 0.1) + (0.35 \times 0.8 \times 0.4) + (0.35 \times 0.2 \times 1)$$

$$= 0.247$$

Shaded on tree

Multiply along the branches for each end result where Don loses the point.

Then add these probabilities.

(iii) P(First serve in | Don loses point)

$$= \frac{\text{P(first serve in and loses point)}}{\text{P(loses point)}}$$

$$= \frac{0.65 \times 0.1}{0.247} \qquad \begin{array}{l} \longleftarrow \text{Marked * on tree.} \\ \longleftarrow \text{Found in part (ii)} \end{array}$$

$$= 0.2631\ldots$$

$$= 0.263 \text{ (3 s.f.)}$$

Example 3.30

In an archery competition, Bill is allowed up to three attempts to hit the target. If he succeeds on any attempt, he does not make any more attempts. The probability that he will hit the target on the first attempt is 0.6. If he misses, the probability that he will hit the target on his second attempt is 0.7. If he misses on the second attempt, the probability that he will hit the target on his third attempt is 0.8.

(i) Draw a fully labelled tree diagram.

(ii) Find the probability that Bill will hit the target.

(iii) Given that Bill hits the target, find the probability that he made at least two attempts.

(i) Let H_n be the event 'Bill hits the target on the nth attempt' and M_n be the event 'Bill misses the target on the nth attempt'.

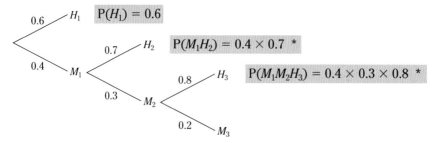

(ii) P(Bill hits target) $= 0.6 + 0.4 \times 0.7 + 0.4 \times 0.3 \times 0.8$
$$= 0.6 + 0.28 + 0.096$$
$$= 0.976$$

(iii) P(at least two attempts | hits target)

$$= \frac{\text{P(at least two attempts and hits target)}}{\text{P(hits target)}}$$

$$= \frac{0.28 + 0.096}{0.976}$$

$$= 0.3852\ldots = 0.385 \text{ (3 s.f.)}$$

Example 3.31

Boxes of sweets contain toffees and chocolates. Box A contains 6 toffees and 4 chocolates, box B contains 5 toffees and 3 chocolates and box C contains 3 toffees and 7 chocolates. One of the boxes is chosen at random and two sweets are taken out, one after the other, and eaten.

(i) Find the probability that they are both toffees.

(ii) Given that they are both toffees, find the probability that they both came from box A.

Cambridge Paper 6 Q2 N05

The tree diagram to illustrate this situation is complicated:

The first set of branches has three outcomes: Box A, Box B, Box C. For each of these, the second set of branches has two outcomes: toffee, chocolate and for each of these the third set of branches also has two outcomes: toffee, chocolate.

So the number of end results $= 3 \times 2 \times 2 = 12$

However, if you keep a clear head, you could find the probabilities without drawing a tree as follows:

Box A has 10 sweets: 6T, 4C

Box B has 8 sweets: 5T, 3C

Box C has 10 sweets: 3T, 7C

(i) P(both toffees)
        = P(Box A, then a toffee, then another toffee)
          + P(Box B, then a toffee, then another toffee)
          + P(Box C, then a toffee, then another toffee)

As the box is chosen at random, each box is equally likely to be chosen,
so $P(\text{box A}) = P(\text{box B}) = P(\text{box C}) = \frac{1}{3}$

$$P(\text{both toffees}) = \frac{1}{3} \times \frac{6}{10} \times \frac{5}{9} + \frac{1}{3} \times \frac{5}{8} \times \frac{4}{7} + \frac{1}{3} \times \frac{3}{10} \times \frac{2}{9}$$
$$= \frac{1}{9} + \frac{5}{42} + \frac{1}{45}$$
$$= \frac{53}{210}$$

(ii) $P(\text{box A} \mid \text{both toffees}) = \dfrac{P(\text{box A and both toffees})}{P(\text{both toffees})}$

$$= \frac{\frac{1}{9}}{\frac{53}{210}}$$
$$= \frac{70}{159}$$

Exercise 3e

1 A coin is biased so that the probability the coin shows heads
when it is tossed is 0.8. The coin is tossed twice.

  (i) Copy and complete the tree diagram, showing the
probabilities on the branches.

  (ii) Find the probability that the coin shows heads both times.

  (iii) Find the probability that the coin shows heads at least once.

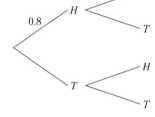

2 The probability that I am late for work on any day is 0.05.

  (i) Copy and compete the tree diagram to show the
outcomes on two days.

  (ii) Find the probability that, on two consecutive days,
I am late for work

    (a) on both days,

    (b) on only one day,

    (c) on the first day, given that I am late for work on only one day.

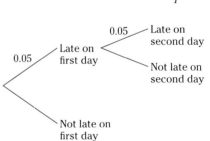

3 A mother and her daughter are both entering a cake competition at a show. From past experience
they estimate that the probability that the mother will win a prize is $\frac{1}{6}$ and, independently of her
mother's result, the probability that the daughter will win a prize is $\frac{2}{7}$.

  (i) Draw a fully labelled tree diagram.

  (ii) Find the probability that

    (a) either the mother or the daughter, but not both, wins a prize,

    (b) the mother wins a prize, given that just one of them wins a prize,

    (c) at least one of them wins a prize.

4 A manufacturer makes pens. When the process is going well, only 2.5% of the pens are defective. Carmela buys two pens. By drawing a tree diagram, or otherwise, find the probability that

    (i) both pens are defective,

    (ii) exactly one of the pens is defective.

5 Two golfers, Chris and Toby, are attempting to qualify for a golf tournament. On past performance, the probability that Toby will qualify is 0.8, the probability that Chris will qualify is p and the probability that **both** Toby and Chris will qualify is 0.6. The event 'Chris qualifies' is independent of the event 'Toby qualifies'.

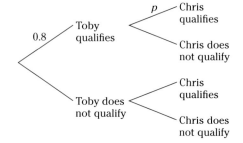

    (i) Find p.

    (ii) Find the probability that just one qualifies.

    (iii) Given that just one qualifies, find the probability that it is Chris.

6 Each day I travel to work by route A or route B. The probability that I choose route A is $\frac{1}{4}$. The probability that I am late for work if I go via route A is $\frac{2}{3}$ and the probability that I am late if I go via route B is $\frac{1}{3}$.

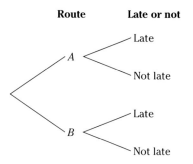

    (i) Find the probability that I am late for work.

    (ii) Given that I am late for work, find the probability that I went via route B.

7 A box contains 20 chocolates, of which 15 have soft centres and 5 have hard centres. Sadie takes two chocolates from the box and eats them.

    (i) Find the probability that just one of the chocolates has a soft centre.

    (ii) Find the probability that the first chocolate has a soft centre, given that just one of the chocolates has a soft centre.

8 In my bookcase there are four shelves. The number of books on each shelf is shown in the table below.

	Hardback	Paperback
Shelf 1	11	9
Shelf 2	8	12
Shelf 3	16	4
Shelf 4	9	3

I choose a shelf at random and then choose a book from the shelf.

    (i) Find the probability that I choose a hardback book.

    (ii) If the book chosen is a hardback, what is the probability that it is from shelf 3?

9 In a large batch of flower seeds 70% have been treated to improve germination. The treated seeds have a probability of 0.8 of germinating, whereas the untreated seeds have a probability of 0.3 of germinating.

A seed is selected at random from the batch.

   (i) Draw a fully labelled tree diagram.

   (ii) Find the probability that the seed will germinate.

   (iii) Find the probability that the seed has been treated, given that it has germinated.

10 Find the probability that a fair cubical die shows 2 on exactly one occasion when it is thrown

   (i) twice,

   (ii) three times.

11 A box contains 9 pens of which 6 are red and 3 are blue. Patrick is doing an experiment as part of his mathematics homework.

   (a) He takes a pen from the box at random, notes its colour and then puts it back in the box. He does this a second time, then a third time.

      Find the probability that he takes out

      (i) a red pen each time,

      (ii) at least one blue pen.

   (b) Patrick now repeats the experiment, but this time he does not return the pen to the box each time.

      Find the probability that he takes out

      (i) a red pen each time,

      (ii) at least one blue pen.

12 Some students are answering multiple choice questions. In each question, there are four choices.

   (a) Amy does not know the answer to a particular question, so she guesses. What is the probability that Amy guesses the correct answer?

   (b) Ben guesses the answers to two of the multiple choice questions. Find the probability that

      (i) both answers are correct,

      (ii) exactly one answer is correct.

   (c) Cathy guesses the answers to three of the multiple choice questions. Find the probability that

      (i) all three are incorrect,

      (ii) exactly two answers are correct,

      (iii) at least two answers are correct,

      (iv) fewer than two answers are correct.

   (d) Darren guesses the answers to four multiple choice questions. Find the probability that all his answers are correct.

13 Pippa is playing a game at a fete in which she randomly selects a coloured disc from a bag. The bag contains 10 discs of which 9 are blue and 1 is white. If she selects the white disc she will win a teddy bear. She is allowed three attempts and no disc is returned to the bag once it has been chosen.

   (i) Find the probability that Pippa wins the teddy bear.

   (ii) Given that she wins the teddy bear, find the probability that she only needed one attempt.

14 Becky is playing table tennis. Each time that she serves, the probability that she wins the point is 0.6, independently of the result of any preceding serves. At the start of a particular game she serves for the first five points.

  (i) Find the probability that, for the first two points of the game,

    (a) she wins both points,

    (b) she wins exactly one of the points.

 (ii) Calculate the probability that for the first five points of the game she loses all five points.

15 Jodie and Kate are playing a game in which they have two bags containing coloured cubes.
Bag A has 7 red cubes and 3 blue cubes.
Bag B has 4 red cubes and 6 blue cubes.

  (i) Jodie takes a cube at random from bag A and Kate takes a cube at random from bag B. Find the probability that

    (a) both cubes are red,

    (b) just one of the cubes is red.

The cubes are returned to their correct bags.

 (ii) Jodie now takes a cube at random from bag A and, without looking at the colour, puts it in bag B. Kate now takes a cube at random from bag B. Find the probability that it is red.

Summary

The **probability** of an event is a measure of the likelihood it will happen.

A probability of 0 indicates that the event is **impossible**.

A probability of 1 (or 100%) indicates that the event is **certain** to happen.

All other events have a probability between 0 and 1.

Equally likely outcomes

$$P(A) = \frac{n(A)}{n(S)}$$ ←—number of outcomes in event A
←—total number of outcomes in the possibility space S

Complement

The complement of A is A' where A' is the event 'A does not occur'.
$P(A') = 1 - P(A)$

Combined events

For events A and B,

$$P(A \text{ or } B) = P(A) + P(B) - P(A \text{ and } B)$$ Remember that '**or**' means A or B or both.

In set notation $P(A \cup B) = P(A) + P(B) - P(A \cap B)$

Mutually exclusive events

For mutually exclusive events A and B

$$P(A \text{ and } B) = 0,$$

so $P(A \text{ or } B) = P(A) + P(B)$ '**or**' rule for mutually exclusive events

In set notation $P(A \cup B) = P(A) + P(B)$

Conditional probability

$$P(A, \text{ given } B) = P(A \mid B) = \frac{P(A \text{ and } B)}{P(B)}$$

so $\qquad\qquad P(A \text{ and } B) = P(B) \times P(A \mid B) = P(A) \times P(B \mid A)$

Independent events

For independent events A and B

$$P(A \mid B) = P(A)$$
$$P(B \mid A) = P(B)$$
$$P(A \text{ and } B) = P(A) \times P(B) \qquad\qquad \textbf{'and'} \text{ rule for independent events}$$

In set notation $\qquad P(A \cap B) = P(A) \times P(B)$

Tree diagrams

For two events A and B, each with two outcomes:

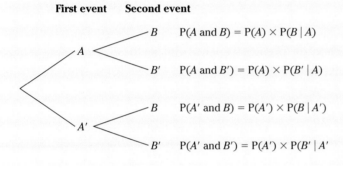

| First event | Second event | | Multiply the probabilities along the branches. |

B $\qquad$ $P(A \text{ and } B) = P(A) \times P(B \mid A)$

A

B' $\qquad$ $P(A \text{ and } B') = P(A) \times P(B' \mid A)$

B $\qquad$ $P(A' \text{ and } B) = P(A') \times P(B \mid A')$

A'

B' $\qquad$ $P(A' \text{ and } B') = P(A') \times P(B' \mid A')$

Mixed Exercise 3

1 On any morning, the probability that I have to wait at the traffic lights on my way to school is $\frac{1}{4}$.

  (i) Find the probability that, on two consecutive mornings, I have to wait at the traffic lights

    (a)   on exactly one morning,

    (b)   on the second morning, given that I have to wait at the traffic lights on exactly one morning,

  (ii) Find the probability that, on three consecutive mornings, I have to wait at least once.

2 A coin is biased so that the probability of obtaining a head is 0.6. The coin is tossed twice.

  (i) Find the probability that at least one head is obtained.

  (ii) Find the probability that exactly one head is obtained, given that at least one head is obtained.

3 A local greengrocer sells fruit, 30% of which is organically grown and 70% is conventionally grown. Sales of apples constitute 20% of the organically grown fruit and 45% of the conventionally grown fruit.

  (i) Show the information on a labelled tree diagram.

  A customer who has purchased fruit is chosen at random to take part in a survey.

  (ii) Find the probability that the customer bought apples.

  (iii) Given that the customer bought apples, find the probability that they were organically grown.

4 A choir has 6 sopranos, 5 altos, 4 tenors and 3 basses. The sopranos and altos are women and the tenors and basses are men. Three members of the choir are chosen at random to give out the programmes before the performance.

  (i) Find the probability that

   (a)   all 3 basses are chosen,

   (b)   2 sopranos and 1 tenor are chosen,

   (c)   3 women are chosen.

  (ii) Given that 3 women are chosen, find the probability that they are all sopranos.

5 The ten letters in the word STATISTICS are arranged at random in a line. Find the probability that the three vowels are together.

6 In a certain country 54% of the population is male. It is known that 5% of the males are colour-blind and 2% of the females are colour-blind. A person is chosen at random and found to be colour-blind. By drawing a tree diagram or otherwise, find the probability that the person is male.

Cambridge Paper 6 Q5 N03

7 Alex and Bonnie play each other at table tennis. Each game results in either a win for Alex or a win for Bonnie. If Alex wins a particular game, the probability of her winning the next game is 0.8, but, if she loses, the probability of her winning the next game is 0.3. The probability of Alex winning the first game is 0.7.

By drawing a tree diagram, or otherwise, find the probability that

  (i) Alex wins the first game, given that she wins exactly one of the first two games,

  (ii) Bonnie loses two games and wins one game in the first three games.

8 A bag contains 7 yellow discs and 3 red discs. Discs are removed at random, one at a time, **without replacement**.

Find the probability that

  (i) the second disc is yellow, given that the first disc was red,

  (ii) the second disc is yellow,

  (iii) the third disc is red, given that the first disc was red.

9 When a particular firm needs to hire a taxi, the receptionist calls one of three firms, X, Y or Z.

40% of the calls are to X, 50% are to Y and 10% are to Z.

9% of the taxis hired from X are late, 6% of those hired from Y are late and 20% of those hired from Z are late.

Find the probability that the next taxi hired

  (i) will be from X and will not arrive late,

  (ii) will arrive late,

  (iii) is from X, given that it arrives late.

10 At a children's party each of the 12 guests is to receive a toy. James is hoping to get a torch and Emma wants a ball. The 12 toys, consisting of 4 balls, 3 torches and 5 pens, are placed in a bag. James receives the first toy drawn out of the bag and Emma receives the second toy drawn out of the bag. Assume that at each stage each toy has an equal chance of being drawn.

(i) Find the probability that Emma will get a ball given that James did not get a torch.

(ii) When Emma's parents arrive at the end of the party, Emma shows them the ball that she got in the draw. Find the probability that James got a torch.

11 Events A and B are such that $P(A) = 0.5$, $P(A \text{ and } B) = 0.2$ and $P(A \text{ or } B) = p$.

Find, in terms of p,

(i) $P(B)$,

(ii) $P(A \text{ given } B)$

If A and B are independent events

(iii) find the value of p.

12 In a group of 12 international referees there are 4 from Asia, 3 from Africa and 5 from Europe. Three referees are chosen at random from the group to officiate at a tournament. Calculate the probability that

(i) the three referees are chosen from the same continent,

(ii) the three referees are all from Africa, given that they are all from the same continent,

(iii) there is one referee from each continent.

13 At a zoo, rides are offered on elephants, camels and jungle tractors. Ravi has money for only one ride. To decide which ride to choose, he tosses a fair coin twice. If he gets 2 heads he will go on the elephant ride. If he gets 2 tails he will go on the camel ride and if he gets 1 of each he will go on the jungle tractor ride.

(i) Find the probabilities that he will go on each of the three rides.

The probabilities that Ravi is frightened on each of the rides are as follows:

elephant ride $\frac{6}{10}$,   camel ride $\frac{7}{10}$,   jungle tractor ride $\frac{8}{10}$.

(ii) Draw a fully labelled tree diagram showing the rides that Ravi could take and whether or not he is frightened.

Ravi goes on a ride.

(iii) Find the probability that he is frightened.

(iv) Given that Ravi is **not** frightened, find the probability that he went on the camel ride.

Cambridge Paper 6 Q5 J09

14 It was found that 68% of passengers on a train used a cell phone during their train journey. Of those using a cell phone, 70% were under 30 years old, 25% were between 30 and 65 years old and the rest were over 65 years old. Of those not using a cell phone, 26% were under 30 years old and 64% were over 65 years old.

(i) Draw a tree diagram to represent this information, giving all the probabilities in decimals.

(ii) Given that one of the passengers is 45 years old, find the probability of this passenger using a cell phone during the journey.

Cambridge Paper 63 Q3 N10

15   The following histogram illustrates the distribution of times, in minutes, that some students spent taking a shower.

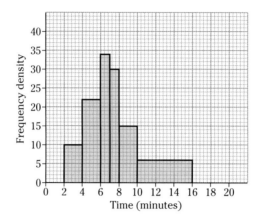

(i) Copy and complete the following frequency table for the data.

Time (*t* minutes)	$2 < t \leqslant 4$	$4 < t \leqslant 6$	$6 < t \leqslant 7$	$7 < t \leqslant 8$	$8 < t \leqslant 10$	$10 < t \leqslant 16$
Frequency						

(ii) Calculate an estimate of the mean time to take a shower.

(iii) Two of these students are chosen at random. Find the probability that exactly one takes between 7 and 10 minutes to take a shower.

Cambridge Paper 63 Q5 N10

16   The lengths of cars travelling on a ferry are noted. The data are summarised in the following table.

Length of car (*x* metres)	Frequency	Frequency density
$2.80 \leqslant x < 3.00$	17	85
$3.00 \leqslant x < 3.10$	24	240
$3.10 \leqslant x < 3.20$	19	190
$3.20 \leqslant x < 3.40$	8	a

(i) Find the value of a.

(ii) Draw a histogram on graph paper to represent the data.

(iii) Find the probability that a randomly chosen car on the ferry is less than 3.20 m in length.

Cambridge Paper 6 Q2 N04

17   Two fair 12-sided dice with sides marked 1, 2, 3, 4, 5, 6, 7, 8, 9, 10, 11, 12 are thrown, and the numbers on the sides which land face down are noted. Events Q and R are defined as follows.

   Q: the product of the two numbers is 24,

   R: both of the numbers are greater than 8.

(i) Find P(Q).

(ii) Find P(R).

(iii) Are events Q and R exclusive? Justify your answer.

(iv) Are events Q and R independent? Justify your answer.

Cambridge Paper 62 Q5 J10

4 Discrete random variables

In this chapter you will learn how to

- construct a probability distribution table for a discrete random variable X
- calculate $E(X)$, the expectation or mean of X
- calculate $Var(X)$, the variance of X

DISCRETE RANDOM VARIABLES

A **variable** is a quantity which may take more than one value. When it is possible to make a list of its individual numerical values, then the variable is said to be **discrete.** An example is the number of coins that show heads when you toss two coins. The possible values are 0, 1 and 2. By contrast a continuous variable such as the height of a four-year old boy cannot be stated precisely. Although you may say that his height is 126 cm, measured to the nearest cm, in practice it could be anywhere in the interval 125.5 cm ⩽ height < 126.5 cm.

A discrete **random** variable is a variable which can take individual values each with a given probability. The values of the variable are usually the outcome of an experiment.

These are some more examples of **discrete random variables**:

	Possible values
The score when you throw an ordinary fair cubical die	1, 2, 3, 4, 5, 6
The number of heads when you toss a fair coin 3 times	0, 1, 2, 3
Your profit in dollars when you play a game with an entry fee of $1 and prizes of $5 and $10	−1, 4, 9
Men's shoe sizes listed in a US catalogue	7.5, 8, 8.5, 9, 9.5, 10, 10.5, 11, 12, 13
The number of times you toss a coin until a tail occurs.	1, 2, 3, 4, ... to infinity

Notation

Consider throwing a fair cubical die. For convenience, an upper case letter, X say, is used as shorthand notation for 'the score on the die'. The values that X can take (i.e. the possible scores) are 1, 2, 3, 4, 5, 6.

The probability that X takes a particular value x is written $P(X = x)$, so the probability that the score is 4, say, is written $P(X = 4)$. Since the die is fair, each number is equally likely to occur so $P(X = 4) = \frac{1}{6}$.

In general:

- Random variables are denoted by upper case letters, for example $X, Y, R, \ldots$
- Particular values of the variable are denoted by lower case letters, for example $x, y, r, \ldots$
- The probability that the variable X takes a particular value x is written $P(X = x)$.
- Alternatively, if the x values are called $x_1, x_2, x_3, \ldots, x_n$, the probabilities can be summarised by writing p_i, where $i = 1, 2, 3, \ldots, n$, so $p_1 = P(X = x_1)$, $p_2 = P(X = x_2)$, and so on.

Probability distributions

A list of all possible values of the discrete random variable X, together with their associated probabilities, is called a **probability distribution**. It is often helpful to show the probability distribution in a table.

The probability distribution of X, the score on the fair cubical die, is shown below, and is illustrated by a vertical line graph.

x	1	2	3	4	5	6
$P(X = x)$	$\frac{1}{6}$	$\frac{1}{6}$	$\frac{1}{6}$	$\frac{1}{6}$	$\frac{1}{6}$	$\frac{1}{6}$

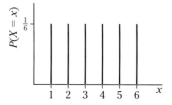

The probability distribution provides a **model** for the scores obtained when the die is thrown a given number of times.

Suppose the die is thrown 120 times. Then the expected frequency for each score is $120 \times \frac{1}{6} = 20$.

When the experiment was performed 120 times the number of times each score occurred was noted. These observed frequencies, together with the expected frequencies according to the probability distribution, are shown below.

Score x	1	2	3	4	5	6
Observed frequency	15	22	23	19	23	18
Expected frequency	20	20	20	20	20	20

Notice that the observed frequencies are close to the expected frequencies, which is what you would expect if the die is fair.

Sum of probabilities

Consider the discrete random variable X with the following probability distribution

x	x_1	x_2	x_3	......	x_n
$P(X = x)$	p_1	p_2	p_3	......	p_n

Since this gives the probabilities of all possible values of the random variable, the sum of all the probabilities is 1.

So $P(X = x_1) + P(X = x_2) + P(X = x_3) + \ldots + P(X = x_n) = 1$

i.e. $\displaystyle\sum_{\text{all } x} P(X = x) = 1$

Alternatively, you could write

$$p_1 + p_2 + p_3 + \ldots + p_n = 1$$

i.e. $\sum p_i = 1$ for $i = 1, 2, 3, \ldots, n$

Often the subscripts are omitted and this is written just as $\sum p = 1$.

Example 4.1

Emma is playing a game with a biased five-sided spinner marked with the numbers 1, 2, 3, 4 and 5.

When she spins the spinner, her score, X, is the number on which the spinner lands. The probability distribution of X is shown in the table.

x	1	2	3	4	5
$P(X = x)$	0.15	0.24	a	0.25	0.19

(i)   Find the value of a.

(ii) Find the probability that the score is at least 4.

(iii) Find the probability that the score is less than 5.

(iv) Find $P(2 < X \leqslant 4)$.

(v)   Write down the most likely score.

(i) $\sum P(X = x) = 1$

   so $0.15 + 0.24 + a + 0.25 + 0.19 = 1$
   $$a + 0.83 = 1$$
   $$a = 0.17$$

(ii) $P(X \text{ is at least } 4)$
   $$= P(X \geqslant 4)$$
   $$= P(X = 4) + P(X = 5)$$
   $$= 0.25 + 0.19$$
   $$= 0.44$$

(iii) $P(X \text{ is less than } 5)$
   $$= P(X < 5)$$
   $$= 1 - P(X = 5)$$
   $$= 1 - 0.19$$
   $$= 0.81$$

   Alternatively,
   $P(X < 5) = P(X = 1) + P(X = 2) + P(X = 3) + P(X = 4)$

   This means that X is greater than 2, but at most 4

(iv) $P(2 < X \leqslant 4)$
   $$= P(X = 3) + P(X = 4)$$
   $$= 0.17 + 0.25$$
   $$= 0.42$$

(v)   The most likely score is the value of X with the greatest probability. This occurs when $X = 4$, so the most likely score is 4.

The most likely score is the **mode** of the distribution.

Example 4.2

George decides to replace the two used batteries in his torch with new ones. Unfortunately, when he takes them out, he mixes them up with three new batteries. All five batteries are identical in appearance.

George selects two of the batteries at random. Draw up a probability distribution table for X, the number of **new** batteries that George selects.

When George selects the batteries he could select no new batteries, 1 new battery or 2 new batteries, so X can take the values 0, 1, and 2.

To find the corresponding probabilities, it is helpful to draw a tree diagram:

Let N be the event 'a new battery is selected'.

Let U be the event 'a used battery is selected'.

The outcomes and probabilities are shown in the tree diagram:

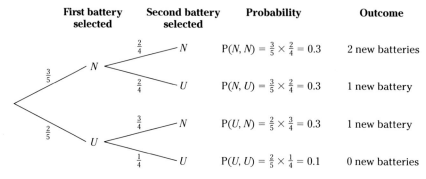

First battery selected	Second battery selected	Probability	Outcome
N ($\frac{3}{5}$)	N ($\frac{2}{4}$)	$P(N, N) = \frac{3}{5} \times \frac{2}{4} = 0.3$	2 new batteries
	U ($\frac{2}{4}$)	$P(N, U) = \frac{3}{5} \times \frac{2}{4} = 0.3$	1 new battery
U ($\frac{2}{5}$)	N ($\frac{3}{4}$)	$P(U, N) = \frac{2}{5} \times \frac{3}{4} = 0.3$	1 new battery
	U ($\frac{1}{4}$)	$P(U, U) = \frac{2}{5} \times \frac{1}{4} = 0.1$	0 new batteries

From the tree diagram,

$P(X = 0) = P(\text{George selects no new batteries}) = 0.1$

$P(X = 1) = P(\text{George selects 1 new battery}) = 0.3 + 0.3 = 0.6$

$P(X = 2) = P(\text{George selects 2 new batteries}) = 0.3$

Probability distribution table:

x	0	1	2
$P(X = x)$	0.1	0.6	0.3

Check the sum of the probabilities:
$P(X = 0) + P(X = 1) + P(X = 2)$
$= 0.1 + 0.6 + 0.3 = 1$, as expected.

Example 4.3

A vegetable basket contains 12 peppers, of which 3 are red, 4 are green and 5 are yellow. Three peppers are taken at random, without replacement, from the basket.

(i) Find the probability that the three peppers are all different colours.

(ii) Show that the probability that exactly 2 of the peppers taken are green is $\frac{12}{55}$.

(iii) The number of **green** peppers taken is denoted by the discrete random variable X. Draw up a probability distribution table for X.

Cambridge Paper 6 Q7 J07

There are 12 peppers: 3R, 4G, 5Y

(i) Number of ways to choose 1R, 1G, 1Y

$$= \binom{3}{1} \times \binom{4}{1} \times \binom{5}{1} = {}_3C_1 \times {}_4C_1 \times {}_5C_1 = 60$$

Total number of ways to choose 3 from 12

$$= \binom{12}{3} = {}_{12}C_3 = 220$$

P(all different colours) $= \dfrac{60}{220} = \dfrac{3}{11}$

Probability involving combinations (page **90**)
Remember $\binom{n}{r} = {}_nC_r$ (page **90**)

(ii) P(exactly 2G and 1 other)

$$= \frac{{}_4C_2 \times {}_8C_1}{{}_{12}C_3} = \frac{48}{220} = \frac{12}{55}$$

(iii) X is the number of green peppers.

X can take the values 0, 1, 2, 3.

$$P(X = 0) = P(0G, 3 \text{ others}) = \frac{{}_8C_3}{{}_{12}C_3} = \frac{14}{55}$$

$$P(X = 1) = P(1G, 2 \text{ others}) = \frac{{}_4C_1 \times {}_8C_2}{{}_{12}C_3} = \frac{28}{55}$$

$$P(X = 2) = P(2G, 1 \text{ other}) = \frac{12}{55} \qquad \text{From part (ii)}$$

$$P(X = 3) = P(3G) = \frac{{}_4C_3}{{}_{12}C_3} = \frac{1}{55}$$

Probability distribution for X:

x	0	1	2	3
$P(X = x)$	$\frac{14}{55}$	$\frac{28}{55}$	$\frac{12}{55}$	$\frac{1}{55}$

Check the sum of the probabilities:
$\sum p = \frac{14}{55} + \frac{28}{55} + \frac{12}{55} + \frac{1}{55} = 1$, as expected.

Probability functions

Sometimes the probability distribution of X can be defined in terms of x by a **probability function.**

Example 4.4

In a probability distribution the random variable X takes the value x with probability kx, where x takes the values 5, 10, 15, 20 and 25 only.

Draw up a probability distribution table for X, in terms of k, and find the value of k.

You are given that $P(X = x) = kx$

So $P(X = 5) = k \times 5 = 5k$

$P(X = 10) = k \times 10 = 10k$, and so on

Probability distribution table for X:

x	5	10	15	20	25
$P(X = x)$	$5k$	$10k$	$15k$	$20k$	$25k$

Since $\sum P(X = x) = 1$

$5k + 10k + 15k + 20k + 25k = 1$

$75k = 1$

$k = \frac{1}{75}$

Example 4.5

Peter has two fair tetrahedral (four-sided) dice. The faces on each die are labelled 1, 2, 3 and 4. One die is red and the other is blue. Peter throws each die once. The random variable X is the sum of the numbers on which the dice land.

  (i) Find the probability that $X = 4$.

  (ii) Draw up a probability distribution table for X.

  (iii) Given that $X = 6$, find the probability that the red die landed on 2.

Show the possible outcomes on a possibility space.

Blue die

		1	2	3	4
	1	2	3	4	5
Red die	2	3	4	5	6
	3	4	5	6	7
	4	5	6	7	8

X can take the values 2, 3, 4, 5, 6, 7 and 8.

Each of the outcomes is equally likely to occur.

  (i) 3 of the 16 outcomes result in a score of 4, so $P(X = 4) = \frac{3}{16}$.

  (ii) Using the diagram to find the probability of each score, the probability distribution is as follows:

x	2	3	4	5	6	7	8
$P(X = x)$	$\frac{1}{16}$	$\frac{2}{16}$	$\frac{3}{16}$	$\frac{4}{16}$	$\frac{3}{16}$	$\frac{2}{16}$	$\frac{1}{16}$

Check the sum of the probabilities:

$\sum P(X = x) = \frac{1}{16}(1 + 2 + 3 + 4 + 3 + 2 + 1) = 1$

  (iii) P(red die landed on 2 | $X = 6$)

$$= \frac{P(\text{red die landed on 2 and } X = 6)}{P(X = 6)} = \frac{\frac{1}{16}}{\frac{3}{16}} = \frac{1}{3}$$

Recall: Conditional probability page **114**

Notice that the probability distribution in Example 4.5 can be described by two functions as follows:

$P(X = x) = \frac{1}{16}(x - 1)$, for $x = 2, 3, 4, 5$

$P(X = x) = \frac{1}{16}(9 - x)$, for $x = 6, 7, 8$

Exercise 4a

1 The discrete random variable X has the probability distribution shown in the table.

x	1	2	3	4	5
$P(X = x)$	0.2	0.25	0.4	a	0.05

  (i) Find the value of a.

  (ii) Draw a vertical line diagram to illustrate the probability distribution.

  (iii) Find

  (a) $P(1 \leqslant X \leqslant 3)$

  (b) the probability that X is at least 3.

  (c) $P(2 < X < 5)$

  (d) $P(X$ is greater than the mode of $X)$

2 The discrete random variable Y has the following probability distribution.

y	10	15	20	25	30
$P(Y = y)$	a	$5a$	$\frac{7}{16}$	$\frac{1}{32}$	$\frac{1}{32}$

  (i) Find the value of a.

  (ii) Find $P(Y > 18)$.

3 The discrete random variable R takes the values 12, 13, 14 and $P(R = r) = kr$, where k is a constant.

  (i) Draw up a probability distribution table in terms of k.

  (ii) Find the value of k.

4 A discrete random variable has probability function

$$P(X = x) = \begin{cases} 0.1 & \text{for } x = -1, 0, 5 \\ a & \text{for } x = 1, 3 \\ 0.3 & \text{for } x = 4 \end{cases}$$

  (i) Write out the probability distribution table in terms of a.

  (ii) Find a.

  (iii) Find $P(X \geqslant 3)$.

5 The discrete random variable X can take the values 0, 1, 2, 3 and 4 only and $P(X = x) = k(x^2 - 2)$.

  (i) Find the value of the constant k.

  (ii) Find $P(1 < X < 4)$.

6 The number of **tails** obtained when two unbiased coins are tossed is denoted by the discrete random variable X.

  (i) Find $P(X = 1)$.

  (ii) Draw up a probability distribution table for X.

7 A drawer contains 8 brown socks and 4 blue socks. Liam takes two socks at random from the drawer, one after the other.

  (i) Show that the probability that Liam takes one brown sock and one blue sock is $\frac{16}{33}$.

  (ii) The discrete random variable B is the number of **brown** socks taken. Draw up a probability distribution table for B.

8 William is playing a game in which he tries to throw tennis balls into a bucket. The probability that the tennis ball lands in the bucket is 0.4 for each attempt.

William has three attempts.

  (i) By drawing a tree diagram, or otherwise, show that the probability that exactly one tennis ball lands in the bucket is 0.432.

  (ii) Draw up a probability distribution table for X, the number of tennis balls that land in the bucket.

William wins a prize if at least two tennis balls land in the bucket.

  (iii) What is the probability that he wins a prize?

9 The discrete random variable X has probability function

$$P(X = x) = \frac{x + 1}{k} \quad \text{for } x = 0, 1, 3, 4.$$

Find the value of the constant k.

10 The discrete random variable X has probability function

$$P(X = x) = \frac{4x - 3}{k} \quad \text{for } x = -1, 0, 1, 2, 3$$

Find the value of the constant k.

11 A fair cubical die has two faces numbered 1, three faces numbered 2 and one face numbered 3. The die is thrown twice. The discrete random variable X is the sum of the two scores.

  (i) Complete the possibility space showing the possible value of X.

		Second throw					
		1	**1**	**2**	**2**	**2**	**3**
First throw	**1**			3			
	1		2				
	2						4
	2					4	
	2	3					
	3						6

  (ii) Draw up a table showing the probability distribution of X.

12 Two fair tetrahedral dice each have faces marked 1, 2, 3 and 4. The two dice are thrown together. The random variable D is zero if both dice land on the same number. If the dice do not land on the same number, then D is the positive difference between the numbers on which they land.

(i) Copy and complete the possibility space.

Second die

		1	**2**	**3**	**4**
First die	**1**			2	
	2	1			
	3				
	4				0

(ii) Draw up a table showing the probability distribution of D.

(iii) Given that $D = 1$, find the probability that one of the dice landed on the face labelled 3.

13 Two ordinary fair cubical dice are thrown. The discrete random variable X is defined as follows. If the two numbers are the same, X is zero. Otherwise X is the smaller of the two numbers.

(i) Show that $P(X = 4) = \frac{1}{9}$.

(ii) Draw up the probability distribution table for X.

E(X), THE EXPECTATION OF X

The **expectation**, or **expected value**, of a random variable X is the result that you would **expect** to get if you took a very large number of values of X and found their mean. It is written E(X) and is denoted by the symbol μ. It is sometimes called the **expected mean**, or just the **mean**.

μ is the Greek letter mu

Experimental approach

Consider again the observed frequency distribution showing the scores on the fair cubical die when it was thrown 120 times.

Score x	1	2	3	4	5	6	
Frequency f	15	22	23	19	23	18	Total $N = 120$

The mean score is $\bar{x}$, where

$$\bar{x} = \frac{\sum x_i f_i}{N} = \frac{1 \times 15 + 2 \times 22 + 3 \times 23 + 4 \times 19 + 5 \times 23 + 6 \times 18}{120} = 3.5583...$$

You could write the formula in a different way as follows:

$$\bar{x} = \sum \left(x_i \times \frac{f_i}{N} \right)$$

$$= x_1 \times \frac{f_1}{N} + x_2 \times \frac{f_2}{N} + ... + x_n \times \frac{f_n}{N}$$

$$= 1 \times \frac{15}{120} + 2 \times \frac{22}{120} + ... + 6 \times \frac{18}{120}$$

$$= 3.5583...$$

The fractions $\frac{15}{120}, \frac{22}{120}, ..., \frac{18}{120}$ are the **relative frequencies** of the scores of 1, 2, ... , 6. Notice that they are quite close to $\frac{20}{120} = \frac{1}{6}$. If you throw the die a large number of times, each of the fractions $\frac{f_i}{N}$ should be very close to $\frac{1}{6}$, the limiting value of the relative frequency of a particular score on the die.

As N becomes very large, the relative frequencies $\frac{f_i}{N}$ tend to the probabilities p_i and $\bar{x}$ tends to the expected mean $\mu = E(X)$.

So, for large values of N,

$$\bar{x} = \frac{\sum x_i f_i}{N} = \sum \left(x_i \times \frac{f_i}{N} \right) \text{ tends to } \mu = E(X) = \sum x_i p_i$$

Theoretical approach

The probability distribution for X, the score on the fair cubical die, is shown below.

x	1	2	3	4	5	6
$P(X = x)$	$\frac{1}{6}$	$\frac{1}{6}$	$\frac{1}{6}$	$\frac{1}{6}$	$\frac{1}{6}$	$\frac{1}{6}$

The expectation, or expected mean, is obtained by multiplying each score by its probability, so

$$\mu = E(X) = 1 \times \tfrac{1}{6} + 2 \times \tfrac{1}{6} + 3 \times \tfrac{1}{6} + 4 \times \tfrac{1}{6} + 5 \times \tfrac{1}{6} + 6 \times \tfrac{1}{6} = 3.5$$

The expectation can be thought of as the average value when the number of experiments increases indefinitely.

If the discrete random variable X has the following probability distribution

x	x_1	x_2	x_3	......	x_n
$P(X = x)$	p_1	p_2	p_3	......	p_n

to calculate the expectation $\mu = E(X)$
- multiply each value x_i by its corresponding probability p_i
- add these products together.

Expectation of the discrete random variable X:

$$\mu = E(X) = x_1 p_1 + x_2 p_2 + x_3 p_3 + \dots + x_n p_n$$
$$= \sum x_i p_i \quad \text{for } i = 1, 2, 3, \dots, n$$

This can also be written   In the examination the formula is given as $E(X) = \sum xp$.

$$\mu = E(X) = \sum x P(X = x)$$

Note
- A **practical** approach results in a **frequency distribution** and an **experimental mean** $\bar{x}$.
- A **theoretical** approach uses a **probability distribution** and results in an **expected mean** μ.

Example 4.6

Natasha plays a fairground game. She throws an unbiased tetrahedral die with faces numbered 1, 2, 3 and 4. If the die lands on the face marked 1 she has to pay \$1. If it lands on 3 she wins 30 cents. If it lands on 2 or 4 Natasha wins 50 cents.

(i)   Find her expected profit in a single throw.

(ii) If the fairground owner changes the rules so that Natasha has to pay \$1.30 if the die lands on 1, what will be Natasha's expected profit in a single throw?

(i) $P(\text{score } 1) = P(\text{score } 2) = P(\text{score } 3) = P(\text{score } 4) = \tfrac{1}{4}$

Let X be Natasha's profit, in cents.

X can take the values -100 (she loses 100 cents), 30 or 50.

$P(X = -100) = P(\text{score } 1) = \tfrac{1}{4}$ As she has to pay 100 cents, her **profit** is negative.

$P(X = 30) = P(\text{score } 3) = \tfrac{1}{4}$

$P(X = 50) = P(\text{score } 2 \text{ or } 4) = \tfrac{1}{4} + \tfrac{1}{4} = \tfrac{1}{2}$

Probability distribution table for X:

x	-100	30	50
$P(X = x)$	$\tfrac{1}{4}$	$\tfrac{1}{4}$	$\tfrac{1}{2}$

$$E(X) = \sum xP(X = x)$$
$$= (-100) \times \tfrac{1}{4} + 30 \times \tfrac{1}{4} + 50 \times \tfrac{1}{2}$$
$$= 7.5$$

Her expected profit in a single throw is 7.5 cents.

(ii) If the die lands on 1, profit $= -130$.

This means that if she played the game 100 times, say, she could expect to make a profit of $7.50.

The new probability distribution table would be as follows:

x	-130	30	50
$P(X = x)$	$\tfrac{1}{4}$	$\tfrac{1}{4}$	$\tfrac{1}{2}$

$$E(X) = \sum xP(X = x)$$
$$= (-130) \times \tfrac{1}{4} + 30 \times \tfrac{1}{4} + 50 \times \tfrac{1}{2}$$
$$= 0$$

Natasha's expected profit in a single throw would now be zero.

Note that if Natasha had to pay **more than** $1.30 whenever the die lands on 1, on average she would lose money, i.e. she would make a loss.

Example 4.7

The discrete random variable X has the following probability distribution.

x	1	3	5	7
$P(X = x)$	0.3	a	b	0.25

(i) Write down an equation satisfied by a and b.
(ii) Given that $E(X) = 4$, find a and b.

Cambridge Paper 6 Q1 N02

(i) *Use the fact that the sum of the probabilities is 1.*

$$\sum P(X = x) = 1$$

So $\quad 0.3 + a + b + 0.25 = 1$
$$a + b + 0.55 = 1$$
$$a + b = 0.45 \qquad (1)$$

(ii) *You have two unknowns, so you will need another relationship between a and b in order to find them.*

Use the fact that $E(X) = 4$.

$$E(X) = \sum xP(X = x)$$

So $\quad 4 = 1 \times 0.3 + 3 \times a + 5 \times b + 7 \times 0.25$
$$4 = 0.3 + 3a + 5b + 1.75$$
$$3a + 5b = 1.95 \qquad (2)$$

Now solve the simultaneous equations.

$$a + b = 0.45 \qquad (1)$$
$$3a + 5b = 1.95 \qquad (2)$$

$(1) \times 3 \qquad 3a + 3b = 1.35 \qquad (3)$

$(2) - (3) \qquad\qquad 2b = 0.6$

$$b = 0.3$$

Sub in (1) $\quad a + 0.3 = 0.45$

$$a = 0.15$$

So, $\quad a = 0.15, b = 0.3$

Example 4.8

In a competition, people pay \$1 to throw a ball at a target. If they hit the target on the first throw, they receive \$5. If they hit it on the second or third throw they receive \$3, and if they hit it on the fourth or fifth throw, they receive \$1. People stop throwing after the first hit, or after 5 throws if no hit is made. Mario has a constant probability of $\frac{1}{5}$ of hitting the target on any throw, independently of the results of other throws.

 (i) Mario misses with his first and second throws and hits the target with his third throw. State how much profit he has made.

 (ii) Show that the probability that Mario's profit is \$0 is 0.184, correct to 3 significant figures.

(iii) Draw up a probability distribution table for Mario's profit.

(iv) Calculate his expected profit.

Cambridge Paper 6 Q6 N05

$P(\text{hit}) = P(H) = \frac{1}{5}, \ P(\text{miss}) = P(M) = \frac{4}{5}, \ \text{Fee} = \$1$

 (i) If he misses with his first and second throws, but hits on his third throw, he receives \$3. He has paid \$1, so he has made \$2 profit.

 (ii) To make a profit of \$0 Mario, having paid \$1, must receive \$1. So he must hit the target on his fourth or fifth attempt but not on any previous attempts.

P(\$0 profit)

$$= P(MMMH) + P(MMMMH)$$

$$= \tfrac{4}{5} \times \tfrac{4}{5} \times \tfrac{4}{5} \times \tfrac{1}{5} + \tfrac{4}{5} \times \tfrac{4}{5} \times \tfrac{4}{5} \times \tfrac{4}{5} \times \tfrac{1}{5}$$

$$= \left(\tfrac{4}{5}\right)^3 \times \tfrac{1}{5} + \left(\tfrac{4}{5}\right)^4 \times \tfrac{1}{5}$$

$$= 0.18432$$

$$= 0.184 \ (3 \text{ s.f.})$$

(iii) *As you have to draw up a probability distribution table you now need to define a random variable X.*

Let X be Mario's profit, in \$.

If Mario hits the target on his first attempt, $X = 5 - 1 = 4$.

$P(X = 4) = P(H) = \frac{1}{5}$

If he hits on his second or third attempt, $X = 3 - 1 = 2$.

$$P(X = 2) = P(MH) + P(MMH)$$
$$= \frac{4}{5} \times \frac{1}{5} + \left(\frac{4}{5}\right)^2 \times \frac{1}{5}$$
$$= 0.288$$

If he hits on his fourth or fifth attempt, $X = 0$.

$P(X = 0) = 0.18432$ from part (ii)

If he misses on all 5 attempts, he receives \$0, so $X = 0 - 1 = -1$

$P(X = -1) = P(MMMMM) = \left(\frac{4}{5}\right)^5 = 0.32768$

Probability distribution table:

x	-1	0	2	4
$P(X = x)$	0.32768	0.18432	0.288	0.2

Check that the probabilities add up to 1.

(iv) $E(X) = \sum xp$
$$= (-1) \times 0.32768 + 0 \times 0.18432 + 2 \times 0.288 + 4 \times 0.2$$
$$= 1.04832 = 1.05 \text{ (3 s.f.)}$$

So Mario's expected profit is \$1.05.

Exercise 4b

1 The probability distribution of the discrete random variable X is shown in the table below.

x	1	2	3	4	5
$P(X = x)$	0.1	0.3	a	0.2	0.1

Find   (i)   the value of a (ii) $E(X)$
        (iii) $P(X > E(X))$.

2 The probability distribution of the discrete random variable Y is shown in the table below.

y	-2	0	1	6
$P(Y = y)$	0.1	0.3	a	0.2

Find the mean of Y.

3 In a probability distribution the random variable X takes the value x with probability kx, where x takes the values 5, 7 and 8 only.

(i) Draw up a probability distribution table for X, in terms of k, and find the value of k.

(ii) Find $E(X)$.

4 Two fair coins are tossed.

(i) Find the probability that exactly one head is obtained.

(ii) Draw up a probability distribution table for X, the number of heads obtained.

(iii) Find the expected number of heads.

5 A bag contains five green counters and six red counters. Two counters are taken at random from the bag, one at a time, and not replaced. The number of red counters taken is denoted by X.

   (i) Draw up a probability distribution table for X.

   (ii) Calculate $E(X)$.

6 The probability distribution of the discrete random variable X is given in the table.

x	10	20	30
$P(X = x)$	a	0.5	b

   (i) Write down an equation satisfied by a and b.

   (ii) Given that $E(X) = 17$, find a and b.

7 Find the expected number of sixes when two fair cubical dice are thrown.

8 A discrete random variable X can take the values 10 and 20 only. The mean of X is 16. Write out the probability distribution of X.

9 The discrete random variable X has probability distribution given by the following, where c is a constant.

$$P(X = x) = \begin{cases} \left(\frac{1}{2}\right)^x & x = 1, 2, 3, 4, 5 \\ c & x = 6 \\ 0 & \text{otherwise} \end{cases}$$

   (i) Find the value of c.

   (ii) Find the mode of X.

   (iii) Find the mean of X.

10 Paula has three keys on a key ring, just one of which opens the door to her house. As she approaches the door, she selects one key after another, at random without replacement, trying each key in the door until she finds the correct key.

   (i) Draw up a probability distribution table for X, the number of keys Paula tries before she opens the door.

   (ii) Calculate the expected number of keys that she will try before opening the door.

11 Every day Eduardo tries to phone his friend. Every time he phones there is a 50% chance that his friend will answer. If his friend answers, Eduardo does not phone again on that day. If his friend does not answer, Eduardo tries again in a few minutes' time. If his friend has not answered after 4 attempts, Eduardo does not try again on that day.

   (i) Draw a tree diagram to illustrate the situation.

   (ii) Let X be the number of unanswered calls made by Eduardo on a day. Copy and complete the table showing the probability distribution of X.

x	0	1	2	3	4
$P(X = x)$		$\frac{1}{4}$			

   (iii) Calculate the expected number of unanswered calls on a day.

Cambridge Paper 6 Q6 J08

VAR(X), THE VARIANCE OF X

The **variance** of a discrete random variable is a measure of the spread of X about the expected mean μ. It is written $Var(X)$ and is denoted by σ^2.

Experimental approach

σ is a Greek letter, read as "sigma".

Recall that the variance of a frequency distribution with mean $\bar{x}$ is given by

$$\text{variance} = \frac{\sum(x_i - \bar{x})^2 f_i}{\sum f_i}$$

$$= \sum(x_i - \bar{x})^2 \times \frac{f_i}{N} \qquad \text{writing } N \text{ for } \sum f_i$$

$$= (x_1 - \bar{x})^2 \times \frac{f_1}{N} + (x_2 - \bar{x})^2 \times \frac{f_2}{N} + \ldots + (x_n - \bar{x})^2 \times \frac{f_n}{N}$$

As n becomes very large,

the relative frequencies $\frac{f_i}{N}$ tend to the probabilities p_i,

$\bar{x}$ tends to μ, the expected mean of X,

and the variance tends to σ^2, the variance of X.

Theoretical approach

The **variance** of a discrete random variable X is defined as follows:

$$\text{Var}(X) = \sigma^2 = \sum (x_i - \mu)^2 p_i \quad \text{for } i = 1, 2, 3, ..., n \quad \text{where } \mu = E(X) = \sum x_i p_i$$

Often the subscripts are omitted

so $\quad \text{Var}(X) = \sum (x - \mu)^2 p$ 　　　　　　　　　　　　　　You could write P($X = x$) instead of p

Note that the **standard deviation** of X, denoted by σ, is the square root of the variance of X,

i.e. $\quad \sigma = \sqrt{\text{Var}(X)}$

Example 4.9

The probability distribution of the discrete random variable X is shown below.
  (i) Find Var(X).
  (ii) Find the standard deviation of X.

x	1	2	3
P($X = x$)	0.3	0.5	0.2

(i) *First find E(X).*

$$E(X) = \mu = \sum xp = 1 \times 0.3 + 2 \times 0.5 + 3 \times 0.2 = 1.9$$

Now apply the variance formula.

$$\begin{aligned}\text{Var}(X) &= \sum (x - 1.9)^2 p \\ &= (1 - 1.9)^2 \times 0.3 + (2 - 1.9)^2 \times 0.5 + (3 - 1.9)^2 \times 0.2 \\ &= 0.49\end{aligned}$$

(ii) standard deviation $= \sqrt{0.49} = 0.7$

Method for finding Var(X)
For each x value:

- Subtract the mean from the x value
- Square the result
- Multiply by the probability

Then add all these products together.

The above formula $\text{Var}(X) = \sum (x_i - \mu)^2 p_i$ can be difficult to use, so an alternative version is usually used instead.

Alternative version of formula for Var(X):

$$\text{Var}(X) = \sigma^2 = \sum x_i^2 p_i - \mu^2 \quad \text{where } \mu = E(X) = \sum x_i p_i \quad i = 1, 2, 3, ..., n$$

In the formula sheet in the examination the subscripts are omitted and the formula is written as

$$\text{Var}(X) = \sum x^2 p - \{E(X)\}^2 \quad \text{with E}(X) \text{ given as } \sum xp$$

The alternative version is derived as follows.   You will NOT need this proof in the examination.

$$\text{Var}(X) = \sum(x_i - \mu)^2 p_i$$

$$= \sum(x_i - \mu)(x_i - \mu)\, p_i$$

$$= \sum(x_i^2 - 2\mu x_i + \mu^2)\, p_i$$

$$= \sum x_i^2 p_i - 2\mu\sum x_i p_i + \mu^2\sum p_i \qquad \sum x_i p_i = \mu \text{ and } \sum p_i = 1$$

$$= \sum x_i^2 p_i - 2\mu^2 + \mu^2$$

$$= \sum x_i^2 p_i - \mu^2$$

To calculate $\text{Var}(X)$ for the discrete random variable X with probability distribution

x	x_1	x_2	x_3	......	x_n
$P(X = x)$	p_1	p_2	p_3	......	p_n

- first calculate $\mu = E(X)$
- now square each x_i and multiply by its corresponding probability p_i to get $x_1^2 p_1$, $x_2^2 p_2$, ...
- add these products together
- then subtract μ^2, i.e. subtract $\{E(X)\}^2$

Example 4.10

The table below shows the probability distribution of the discrete random variable X.

x	1	2	3	4
$P(X = x)$	0.1	0.3	0.45	0.15

(i) Calculate $E(X)$.
(ii) Calculate $\text{Var}(X)$.
(iii) Find the standard deviation of X.

(i) $\mu = E(X) = \sum xp$

$$= 1 \times 0.1 + 2 \times 0.3 + 3 \times 0.45 + 4 \times 0.15$$

$$= 2.65$$

(ii) $\text{Var}(X) = \sum x^2 p - \mu^2$

$$= 1^2 \times 0.1 + 2^2 \times 0.3 + 3^2 \times 0.45 + 4^2 \times 0.15 - (2.65)^2$$

$$= 0.7275$$

Remember that the x-values are squared but the probabilities are not.

Remember to subtract μ^2 at the end.

(iii) Standard deviation $= \sqrt{0.7275}$

$$= 0.8529...$$

$$= 0.853 \ (3 \text{ s.f.})$$

Example 4.11

A small farm has 5 ducks and 2 geese. Four of these birds are to be chosen at random. The random variable X represents the number of geese chosen.

(i) Draw up the probability distribution of X.

(ii) Show that $E(X) = \frac{8}{7}$ and calculate $\text{Var}(X)$.

Cambridge Paper 62 Q6(i) and (ii) J10

(i) X is the number of geese chosen and X can take the values 0, 1 and 2.

Total number of ways to choose 4 birds from 7 birds with no restrictions

$$= \binom{7}{4} = {}_7C_4 = 35$$

Consider the different numbers of geese that could be chosen and calculate the probability for each.

Number of ways to choose 0 geese (i.e. 4 ducks) $= \binom{5}{4} = 5$

So $P(X = 0) = \frac{5}{35} = \frac{1}{7}$

Number of ways to choose 1 goose and 3 ducks $= \binom{2}{1} \times \binom{5}{3} = 20$

So $P(X = 1) = \frac{20}{35} = \frac{4}{7}$

Number of ways to choose 2 geese and 2 ducks $= 1 \times \binom{5}{2} = 10$

So $P(X = 2) = \frac{10}{35} = \frac{2}{7}$

Probability distribution:

x	0	1	2
$P(X = x)$	$\frac{1}{7}$	$\frac{4}{7}$	$\frac{2}{7}$

Remember to check that the sum of the probabilities is 1.

(ii) $E(X) = \sum xp$

$$= 0 \times \tfrac{1}{7} + 1 \times \tfrac{4}{7} + 2 \times \tfrac{2}{7}$$

$$= \tfrac{8}{7}, \text{ as required. } \mu^2$$

$$\text{Var}(X) = \sum x^2 p - \{E(X)\}^2$$

$$= 0^2 \times \tfrac{1}{7} + 1^2 \times \tfrac{4}{7} + 2^2 \times \tfrac{2}{7} - \left(\tfrac{8}{7}\right)^2$$

$$= \tfrac{20}{49}$$

Exercise 4c

1 The discrete random variable X has the following probability distribution.

x	3	5	7	9
$P(X = x)$	0.2	0.3	0.4	0.1

Calculate (i) the mean of X

(ii) the variance of X.

2 The discrete random variable R has the following probability distribution.

r	-2	-1	0	1	2
$P(R = r)$	0.05	a	0.43	$3a$	0.12

(i) Find the value of a.

(ii) Find $E(R)$ and $\text{Var}(R)$.

3 Find Var(X) for each of the following probability distributions.

(i)

x	-3	-2	0	2	3
P($X = x$)	0.3	0.3	0.2	0.1	0.1

(ii)

x	1	3	5	7	9
P($X = x$)	$\frac{1}{6}$	$\frac{1}{4}$	$\frac{1}{6}$	$\frac{1}{4}$	$\frac{1}{6}$

(iii)

x	0	2	5	6
P($X = x$)	0.11	0.35	0.46	0.08

4 Two boxes each contain three cards. The first box contains cards labelled 1, 3 and 5. The second box contains cards labelled 2, 6 and 8. In a game, a player picks a card at random from each box. The score, X, is the sum of the numbers on the two cards.

(i) By using a possibility space, or otherwise, list the six possible values of X and calculate the corresponding probabilities.

(ii) Calculate the expected score.

(iii) Calculate the variance of X.

(iv) Calculate the standard deviation of X.

5 A computer is programmed to produce a sequence of integers, X, in the range 0 to 5 inclusive, with probabilities as shown in the probability distribution table.

x	0	1	2	3	4	5
P($X = x$)	k	$\frac{1}{30}$	$\frac{2}{30}$	$\frac{3}{30}$	$\frac{4}{30}$	$\frac{5}{30}$

(i) Show that $k = \frac{1}{2}$.

(ii) Calculate E(X).

(iii) Calculate Var(X).

6 A class consists of 7 girls and 5 boys. Three students from the class are chosen at random to represent the class. The number of boys chosen is denoted by the random variable X.

(i) Show that P($X = 2$) $= \frac{7}{22}$.

(ii) Draw up the probability distribution of X.

(iii) Calculate E(X).

(iv) Calculate the variance of X.

7 A discrete random variable X has the following probability distribution and can only take the values tabulated. The mean of X is 6.

x	1	3	6	n	12
P($X = x$)	0.1	0.3	k	0.25	0.15

(i) Find the value of k.

(ii) Find the value of n.

(iii) Find the variance of X.

8 The probability distribution of the random variable X is shown in the following table.

x	2	4	6	8	10
P($X = x$)	0.06	a	b	b	0.16

The mean is 6.28.

(i) Write down two equations involving a and b and hence find the values of a and b.

(ii) Calculate the variance of X.

9 The random variable X takes values 2, 4, 6 and 8. Its probability distribution is illustrated in the vertical line graph.

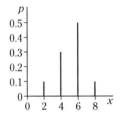

Find Var(X).

10 Anne plays a game in which an unbiased cubical die is thrown once. If the score is 1, 2 or 3 Anne loses $10. If the score is 4 or 5, Anne wins $x. If the score is 6, Anne wins $2x.

(i) Show that the expected value of Anne's profit in a single game is $\$\left(\frac{2}{3}x - 5\right)$.

(ii) Calculate the value of x for which, on average, Anne's profit is zero.

(iii) Given that $x = 12$, calculate the variance of Anne's profit in a single game.

11 The discrete random variable X has probability distribution given by the following, where k is a constant.

$$P(X = x) = \begin{cases} \dfrac{kx}{(x^2 + 1)} & x = 2, 3 \\ \dfrac{2kx}{(x^2 - 1)} & x = 4, 5 \end{cases}$$

(i) Show that $k = \frac{20}{33}$.

(ii) Find the probability that X is less than 3 or greater than 4.

(iii) Find E(X).

(iv) Find Var(X).

Summary

A list of all possible values of the discrete random variable X, together with their associated probabilities, is called a **probability distribution**.

The **sum** of the probabilities of all possible values of a discrete random variable X is 1.

i.e. $\sum P(X = x) = 1$ or $\sum p = 1$ p is shorthand for $P(X = x)$

$E(X)$, the expectation (mean, expected value) of X

$$\mu = E(X) = \sum xp$$

$Var(X)$, the variance of X

$$\sigma^2 = Var(X) = \sum (x - \mu)^2 p$$

Alternative version

$$\sigma^2 = Var(X) = \sum x^2 p - \{E(X)\}^2 \quad \text{i.e. } Var(X) = \sum x^2 p - \mu^2$$

Mixed Exercise 4

1 A box contains 10 pens of which 3 are new. A random sample of two pens is taken.

   (i) Show that the probability of getting exactly one new pen in the sample is $\frac{7}{15}$.

   (ii) Construct a probability distribution table for the number of new pens in the sample.

   (iii) Calculate the expected number of new pens in the sample.

                    Cambridge Paper 6 Q2 J03

2 A fair tetrahedral die has four faces numbered 1, 2, 3 and 6. The die and a fair coin are tossed together. If the coin shows heads the score S is equal to double the number on the hidden face of the die. If the coin shows tails then the score S is equal to the number on the hidden face of the die.

   (i) Copy and complete the probability space showing the possible outcomes

                       Die

		1	2	3	6
Coin	H				6
	T		2		

   (ii) Show that $P(S = 6) = \frac{1}{4}$.

   (iii) Draw up the probability distribution table for S.

   (iv) Show that the expected value of S is 4.5.

   (v) Calculate the variance of S.

3 The discrete random variable D is the number of parcels delivered in a day to a particular house. The probability distribution for D is shown below.

d	0	1	2	3	4
$P(D = d)$	0.1	0.4	0.25	0.15	a

   (i) Find the value of a.

   (ii) Write down the most likely number of parcels delivered.

   (iii) Find the mean number of parcels delivered.

   (iv) Find the probability that the number of parcels delivered is fewer than the mean.

   (v) Find the variance of D.

4 The number of spelling mistakes, X, that Jo makes when writing an essay can be modelled by the following probability distribution.

x	0	1	2	3	4
$P(X = x)$	0.23	0.31	0.27	0.14	0.05

The mean number of spelling mistakes is μ and the standard deviation is σ.

   (i) Find the values of μ and σ.

   (ii) Find $P(X > \mu + \sigma)$.

5 A discrete random variable X has the following probability distribution.

x	1	2	3	4
$P(X = x)$	$3c$	$4c$	$5c$	$6c$

(i) Find the value of the constant c.

(ii) Find $E(X)$ and $Var(X)$.

(iii) Find $P(X > E(X))$.

Cambridge Paper 6 Q8 N03

6 A box contains five balls numbered 1, 2, 3, 4, 5. Three balls are drawn randomly at the same time from the box.

(i) By listing all possible outcomes (123, 124, etc) find the probability that the sum of the three numbers drawn is an odd number.

The random variable L denotes the largest of the three numbers drawn.

(ii) Find the probability that L is 4.

(iii) Draw up a table to show the probability distribution of L.

(iv) Calculate the expectation and variance of L.

Cambridge Paper 6 Q6 N04

7 The probability of there being X unusable matches in a box is given by $P(X = 0) = 8k$, $P(X = 1) = 5k$, $P(X = 2) = P(X = 3) = k$, $P(X \geqslant 4) = 0$.

(i) Find the value of the constant k.

(ii) Find $P(X < E(X))$.

(iii) Find the variance of X.

8 The discrete random variable X has the following probability distribution.

x	0	1	2	3	4
$P(X = x)$	0.26	q	$3q$	0.05	0.09

(i) Find the value of q.

(ii) Find $E(X)$ and $Var(X)$.

Cambridge Paper 6 Q2 N06

9 A bag contains 300 discs of different colours. There are 100 pink discs, 100 blue discs and 100 orange discs. The discs of each colour are numbered from 0 to 99. Five discs are selected at random, one at a time, with replacement. Find

(i) the probability that no orange discs are selected,

(ii) the probability that exactly 2 discs with numbers ending in a 6 are selected,

(iii) the probability that exactly 2 orange discs with numbers ending in a 6 are selected,

(iv) the mean and variance of the number of pink discs selected.

Cambridge Paper 6 Q5 N05

10 The probability distribution of the random variable X is shown in the following table.

x	-2	-1	0	1	2	3
$P(X = x)$	0.08	p	0.12	0.16	q	0.22

The mean of X is 1.05.

(i) Write down two equations involving p and q and hence find the values of p and q.

(ii) Find the variance of X.

Cambridge Paper 61 Q2 N09

11 Box A contains 5 red paper clips and 1 white paper clip. Box B contains 7 red paper clips and 2 white paper clips. One paper clip is taken at random from Box A and transferred to Box B. One paper clip is then taken at random from Box B.

(i) Find the probability of taking both a white paper clip from Box A and a red paper clip from Box B.

(ii) Find the probability that the paper clip taken from Box B is red.

(iii) Find the probability that the paper clip taken from Box A was red, given that the paper clip taken from Box B is red.

(iv) The random variable X denotes the number of times that a red paper clip is taken. Draw up a table to show the probability distribution of X.

Cambridge Paper 6 Q7 N07

5 The binomial distribution

In this chapter you will learn

- the conditions for a discrete random variable to follow a binomial distribution
- how to find binomial probabilities
- how to find the expectation and variance of a binomial variable

THE BINOMIAL DISTRIBUTION

The binomial distribution is the distribution of a special discrete random variable. The following example illustrates how a binomial situation arises.

Example 5.1

A coin is biased so that the probability of obtaining a head when the coin is tossed is 0.7. Find the probability that exactly 2 heads are obtained when the coin is tossed

  (i) 3 times,

  (ii) 6 times.

  (i) Let X be the number of heads in 3 tosses.

To find the probability of obtaining exactly 2 heads in 3 tosses   Tree diagrams page **123**
- *Draw a tree diagram with three sets of branches, one for each toss of the coin.*
- *Locate the end results that give exactly 2 heads, i.e. 2 heads and 1 tail.*
- *Find the sum of the probabilities of these end results.*

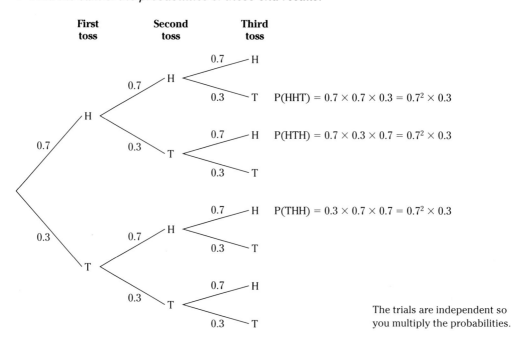

P(HHT) $= 0.7 \times 0.7 \times 0.3 = 0.7^2 \times 0.3$

P(HTH) $= 0.7 \times 0.3 \times 0.7 = 0.7^2 \times 0.3$

P(THH) $= 0.3 \times 0.7 \times 0.7 = 0.7^2 \times 0.3$

The trials are independent so you multiply the probabilities.

P(exactly 2 heads in 3 tosses)

$$= P(X = 2)$$
$$= P(\text{HHT}) + P(\text{HTH}) + P(\text{THH})$$
$$= \mathbf{3} \times 0.7^2 \times 0.3$$
$$= 0.441$$

There are **3** ways of arranging 2 heads and 1 tail.

(ii) Now let X be the number of heads in **6** tosses.

You could extend the tree so that there are six sets of branches but this would be very time-consuming. Instead, just focus on the end results giving exactly 2 heads. These come from all the different arrangements of 2 heads and 4 tails.

One arrangement of 2 heads and 4 tails is HHTTTT in this order.

$$P(\text{HHTTTT}) = 0.7 \times 0.7 \times 0.3 \times 0.3 \times 0.3 \times 0.3$$
$$= 0.7^2 \times 0.3^4$$

But there are several other arrangements such as HTHTTT, TTTHTH, TTHTHT, ... each with probability $0.7^2 \times 0.3^4$.

How many arrangements are there?

The number of arrangements with 2 heads and 4 tails is the same as the number of different ways to **choose** 2 spaces from 6 spaces for H and then fill the remaining spaces with T.

Combinations, page **90**.

The number of ways to choose 2 from 6

$$= \binom{6}{2} = {}_6C_2 = \frac{6!}{2!4!} = 15.$$

On calculator [6] [$_nC_r$] [2] [=]

So P(X = 2)

$$= \binom{6}{2} \times 0.7^2 \times 0.3^4 = \mathbf{15} \times 0.7^2 \times 0.3^4 = 0.0595 \text{ (3 s.f.)}$$

There are **15** ways of arranging 2 heads and 4 tails.

6 choose 2   2 heads   4 tails

Note that the indices add up to 6.

A similar method can be used to find the probability of *any* number of heads in any finite number of tosses. For example,

P(exactly 10 heads in 16 tosses)

$$= P(X = 10)$$
$$= \binom{16}{10} \times 0.7^{10} \times 0.3^6 = \mathbf{8008} \times 0.7^{10} \times 0.3^6 = 0.165 \text{ (3 s.f.)}$$

16 choose 10   10 heads   6 tails

There are **8008** ways of arranging 10 heads and 6 tails.

The indices add up to 16.

The above are examples of a **binomial situation**:

The number of tosses is fixed and the outcomes of the tosses are independent. Also, for each toss, the probability of a head (success) is constant at 0.7 and the probability of a tail (failure) is 0.3.

The general theory for the binomial distribution is summarised below.

Conditions for a binomial distribution

A discrete random variable X follows a **binomial distribution** when all the following conditions are satisfied.
- There are a fixed number of repeated trials.
- The trials are independent.
- Each trial results in one of two outcomes: success or failure.
- The probability of success, p, is constant for each trial.

Note that probability of failure, q, is also constant and, since $p + q = 1$, $q = 1 - p$.

X is defined as the **number of successful outcomes** in n trials.

The distribution of X is written

$$X \sim \mathrm{B}(n, p)$$

Only the values of n and p are needed to describe the distribution fully. They are the **parameters** of the distribution.

This is read as 'X follows a binomial distribution. There are n trials, and the probability of a successful outcome in each trial is p.'

Calculating binomial probabilities

The probability of r successes in n trials is

$$\mathrm{P}(X = r) = \binom{n}{r}p^r q^{n-r} \quad \text{for } r = 0, 1, 2, \ldots, n \quad \text{where } q = 1 - p$$

Recall $\binom{n}{r} = {}_nC_r = \dfrac{n!}{r!(n-r)!}$

Number of ways to choose r from n Probability of r successes   Probability of $(n-r)$ failures

Note that in the list of formulae provided in the examination, the probability is written

$$p_r = \binom{n}{r}p^r(1 - p)^{n-r}$$

Example 5.2

At the Sell-it-all supermarket 60% of customers pay by credit card. Find the probability that in a randomly selected sample of 12 customers

  (i)   exactly 7 pay by credit card,

  (ii) at least 3 but fewer than 5 pay by credit card.

Define 'paying by credit card' as success and 'not paying by credit card' as failure.

Let X be the number of customers in a sample of 12 who pay by credit card.

Define the variable in words.

Assuming that a customer's method of payment is independent of the method used by any other customer, X has a binomial distribution.

$X \sim \mathrm{B}(12, 0.6)$ with $n = 12$, $p = 0.6$, $q = 1 - p = 0.4$

State how the variable is distributed.

  (i)   P(exactly 7 pay by credit card)

$$= \mathrm{P}(X = 7)$$

$$= \binom{12}{7} \times 0.6^7 \times 0.4^5$$

$\binom{12}{7} = {}_{12}C_7 = 792$

$$= 0.2270\ldots$$

$$= 0.227 \text{ (3 s.f.)}$$

  (ii) P(at least 3 but fewer than 5 pay by credit card)

$$= \mathrm{P}(3 \leqslant X < 5)$$

$$= \mathrm{P}(X = 3) + \mathrm{P}(X = 4)$$

$$= \binom{12}{3} \times 0.6^3 \times 0.4^9 + \binom{12}{4} \times 0.6^4 \times 0.4^8$$

$\binom{12}{3} = {}_{12}C_3 = 220$

$\binom{12}{4} = {}_{12}C_4 = 495$

$$= 0.05449\ldots$$

$$= 0.0545 \text{ (3 s.f.)}$$

Example 5.3

The random variable X has distribution $B(7, \frac{1}{5})$. Find

(i) $P(X \leq 2)$,   (ii) $P(X \geq 6)$.

$X \sim B\left(7, \frac{1}{5}\right)$ with $n = 7$, $p = \frac{1}{5}$, $q = 1 - p = \frac{4}{5}$

(i) $P(X \leq 2)$

$= P(X = 0) + P(X = 1) + P(X = 2)$

$= \binom{7}{0} \times \left(\frac{1}{5}\right)^0 \times \left(\frac{4}{5}\right)^7 + \binom{7}{1} \times \left(\frac{1}{5}\right)^1 \times \left(\frac{4}{5}\right)^6 + \binom{7}{2} \times \left(\frac{1}{5}\right)^2 \times \left(\frac{4}{5}\right)^5$

$\binom{7}{0} = 1$ and $\left(\frac{1}{5}\right)^0 = 1$, so you could just write $P(X = 0)$ as $\left(\frac{4}{5}\right)^7$.

$= 0.2097... + 0.3670... + 0.2752...$

$= 0.8519...$

$= 0.852$ (3 s.f.)

(ii) $P(X \geq 6)$

$= P(X = 6) + P(X = 7)$

$= \binom{7}{6} \times \left(\frac{1}{5}\right)^6 \times \left(\frac{4}{5}\right)^1 + \binom{7}{7} \times \left(\frac{1}{5}\right)^7 \times \left(\frac{4}{5}\right)^0$

$\binom{7}{7} = 1$ and $\left(\frac{4}{5}\right)^0 = 1$, so you could just write $P(X = 7)$ as $\left(\frac{1}{5}\right)^7$.

$= 0.0003584 + 0.0000128$

$= 0.0003712$

$= 0.000371$ (3 s.f.)

You may be able to shorten the working in some questions by using the following:

$P(\text{no successes}) = P(X = 0) = q^n$

$P(\text{all successes}) = P(X = n) = p^n$

Example 5.4

On a certain road 20% of the vehicles are trucks, 16% are buses and the remainder are cars. A random sample of 11 vehicles is taken. Find the probability that fewer than 3 are buses.

Cambridge Paper 6 Q3(i) J09

At first sight there appear to be three possible outcomes for each trial: the vehicle is a bus, truck or car.

*However you are interested only in the number of buses, so consider 'the vehicle is a bus' as success and 'the vehicle is **not** a bus' as failure.*

Let X be the number of buses in a sample of 11 vehicles.

$X \sim B(11, 0.16)$ with $n = 11$, $p = 16\% = 0.16$, $q = 1 - p = 0.84$

$P(\text{fewer than 3 are buses})$

$= P(X < 3)$

$= P(X = 0) + P(X = 1) + P(X = 2)$

$= 0.84^{11} + \binom{11}{1} \times 0.16^1 \times 0.84^{10} + \binom{11}{2} \times 0.16^2 \times 0.84^9$

$= 0.1469... + 0.3078... + 0.2931...$

$= 0.7479...$

$= 0.748$ (3 s.f.)

Exercise 5a

1 If $X \sim B(10, 0.3)$ find
    (i) $P(X = 2)$ (ii) $P(X = 6)$
    (iii) $P(X = 10)$ (iv) $P(X = 0)$

2 If $X \sim B(8, 0.25)$ find
    (i) $P(X = 5)$ (ii) $P(X = 3)$
    (iii) $P(X \leqslant 3)$ (iv) $P(X \geqslant 7)$

3 If $X \sim B(9, \frac{1}{3})$ find
    (i) $P(X = 5)$
    (ii) $P(X < 2)$
    (iii) $P(X > 7)$

4 If $X \sim B(9, 0.45)$ find
    (i) $P(2 \leqslant X \leqslant 4)$
    (ii) $P(5 < X < 8)$

5 If $X \sim B(13, 0.7)$ find
    (i) $P(8 < X \leqslant 10)$
    (ii) $P(8 \leqslant X < 10)$

6 A fair coin is tossed 12 times. Find the probability that the number of tails obtained is
    (i) exactly 6,
    (ii) at least 10,
    (iii) no more than 2.

7 The probability that a patient attending a clinic has a particular health condition is $\frac{2}{5}$. Find the probability that in a randomly chosen group of 7 patients attending the clinic
    (i) exactly 3 have the condition,
    (ii) more than 5 have the condition,
    (iii) fewer than 2 have the condition,
    (iv) at least 2 but no more than 4 have the condition.

8 A fair tetrahedral die with faces numbered 1, 2, 3 and 4 is thrown 10 times. Find the probability of obtaining
    (i) exactly 4 fours,
    (ii) fewer than 3 odd numbers.

9 In a particular region 10% of people have blood type B.
    (i) Find the probability that exactly 3 have blood type B in a random sample of 5 people from the region.
    (ii) Find the probability that at most 2 have blood type B in a random sample of 9 people from the region.
    (iii) Find the probability that exactly 13 do **not** have blood type B in a random sample of 15 people from the region.

10 In a survey it is found that 48% of pupils travel to the local school by bus. Find the probability that, in a random sample of 6 pupils, more than half of the pupils travel to school by bus.

11 A bag contains counters, 35% of which are red and the rest are yellow. Alex selects a counter at random from the bag, notes its colour and puts it back into the bag. Alex does this 8 times in all. Find the probability that Alex picks
    (i) exactly 3 red counters,
    (ii) more than 6 yellow counters.

12 In an experiment 5 fair coins are tossed together. Find the probability that they land showing
    (i) exactly 3 tails,
    (ii) fewer than 3 tails,
    (iii) more than 3 heads.

USEFUL METHODS

Sum of probabilities

Recall that, for any discrete random variable X, the sum of all the probabilities is 1.

i.e. $$\sum_{\text{all } x} P(X = x) = \sum p = 1 \qquad\qquad \text{See page } \mathbf{140}$$

Using this can shorten the working in some binomial questions, as in the following example.

Example 5.5

On average, three quarters of the patients who have a check-up at a particular dental practice do not need follow-up treatment. Find the probability that, in a random sample of 9 patients from the practice, at most 7 do not need follow-up treatment.

Let X be the number of patients in a sample of 9 who do not need follow-up treatment.

$X \sim B\left(9, \frac{3}{4}\right)$ with $n = 9, p = \frac{3}{4}, q = 1 - p = \frac{1}{4}$

P(at most 7 do not need follow-up treatment)

$= P(X \leqslant 7)$

$= P(X = 0) + P(X = 1) + P(X = 2) + ... + P(X = 7)$

This involves a lot of calculation, so use the fact that the sum of all the probabilities is 1.

$P(X \leqslant 7)$

$= 1 - P(X > 7)$

$= 1 - (P(X = 8) + P(X = 9))$

$= 1 - \left(\binom{9}{8} \times \left(\frac{3}{4}\right)^8 \times \left(\frac{1}{4}\right)^1 + \left(\frac{3}{4}\right)^9\right)$

$= 1 - 0.3003...$

$= 0.69966...$

$= 0.700$ (3 s.f.)

Example 5.6

A box contains a large number of pens and for each pen the probability that it is faulty is 0.1. Laura selects n pens at random from the box. What is the minimum value of n for which the probability she selects at least one faulty pen is greater than 0.95?

Let X be the number of faulty pens when n pens are selected from the box.

So $X \sim B(n, 0.1)$ with n unknown, $p = 0.1, q = 1 - p = 0.9$ You have to find n.

P(at least one faulty pen) $= P(X \geqslant 1)$.

You want $P(X \geqslant 1) > 0.95$.

Now $P(X \geqslant 1) = 1 - P(X = 0)$ $P(X = 0) = q^n$

$= 1 - 0.9^n$

So you want $1 - 0.9^n > 0.95$

Rearranging $0.05 > 0.9^n$

i.e. $0.9^n < 0.05$

Using trial and improvement on your calculator, you will find that

$0.9^{28} = 0.0523... > 0.05$

$0.9^{29} = 0.0471... < 0.05$

So the minimum value of n is **29**.

Alternatively, it is possible to find n using logarithms. Although not in P1, some students may have studied this theory.

You will **not** be required to use logarithms in S1.

Starting with

$$0.9^n < 0.05$$

take logs to base 10 of both sides to get

$$n \log 0.9 < \log 0.05$$

From the calculator you find that log 0.9 = -0.045, so when you divide both sides by log 0.9 you are dividing by a negative quantity. To do this you must reverse the inequality.

So $\qquad n > \dfrac{\log 0.05}{\log 0.9}$

$\qquad\qquad n > 28.4\ldots$ *n* must be a whole number.

The minimum value of *n* is 29, as before.

Two-stage questions

You may need to use the binomial distribution to find a probability which then becomes the probability of success ('*p*') in a second binomial situation. This is illustrated in the following example.

Example 5.7

The Fair Choc Company makes small chocolate eggs, 48% of which are milk chocolate and the remainder of which are plain chocolate. The eggs are mixed before being put into identical foil wrappings and placed at random into boxes, each containing 12 eggs.

  (i) A box is chosen at random. Show that the probability that this box contains exactly 6 milk chocolate eggs is 0.223 correct to 3 significant figures.

 (ii) The manager takes a random sample of 10 boxes of eggs from the production line. Find the probability that at least two of these boxes contain exactly 6 milk chocolate eggs.

  (i) Let X be the number of **milk chocolate eggs** in a **box of 12 chocolate eggs**.

$\qquad X \sim \text{B}(12, 0.48)$ with $n = 12, p = 0.48, q = 1 - p = 0.52$

$\qquad \text{P}(X = 6) = \dbinom{12}{6} \times 0.48^6 \times 0.52^6$

$\qquad\qquad\qquad = 0.2234\ldots$ ⟵ Store in your calculator memory

$\qquad\qquad\qquad = 0.223$ (3 s.f.)

 (ii) Let Y be the number of **boxes that contain exactly 6 milk** You need to define a new variable.
chocolate eggs in a random sample of **10 boxes**.

$\qquad Y \sim \text{B}(10, 0.2234\ldots)$ with $n = 10, p = 0.2234\ldots,$
$\qquad q = 1 - p = 0.7765\ldots$

Use the answer stored in the calculator memory from part (i) as the value of *p* in part (ii).

$\qquad \text{P}(Y \geqslant 2)$

$\qquad\qquad = 1 - \text{P}(Y < 2)$

$\qquad\qquad = 1 - \left((0.7765\ldots)^{10} + \dbinom{10}{1} \times (0.2234\ldots)^1 \times (0.7765\ldots)^9 \right)$

$\qquad\qquad = 1 - 0.30925\ldots$

$\qquad\qquad = 0.69074\ldots$

$\qquad\qquad = 0.691$ (3 s.f.)

Calculator note about accuracy

In the above example, a value calculated in part (i) is needed in part (ii). In this situation you must take care not to lose accuracy. If your final answer is required to 3 significant figures then you must work to at least 4 figures throughout your calculation. Ideally you should store the value in your calculator and then recall it as necessary.

Rounding too early is known as **premature rounding** and this is a very common cause of **loss of marks** in the examination.

For example, suppose you use your rounded answer from part (i) in part (ii), i.e. $p = 0.223$.

Your calculations would then be as follows:

> (ii) $Y \sim B(10, 0.223)$ with $n = 10, p = 0.223, q = 0.777$
>
> $P(Y \geqslant 2)$
>
> $\quad = 1 - P(Y < 2)$
>
> $\quad = 1 - (0.777^{10} + \binom{10}{1} \times 0.223^1 \times 0.777^9)$
>
> $\quad = 1 - 0.3104\ldots$
>
> $\quad = 0.6895\ldots$
>
> $\quad = 0.690 \text{ (3 s.f.)}$

This answer is **not** accurate to 3 significant figures.

Link with binomial expansion of $(q + p)^n$

There is a link between the probabilities in the **binomial distribution** and the terms in the **binomial expansion** of $(q + p)^n$ studied in P1.

Consider $X \sim B(5, p)$. The probability distribution of X is as follows:

See probability distributions page **140**.

$P(X = 0) = \binom{5}{0} p^0 q^5 = q^5$ $q = 1 - p$

$P(X = 1) = \binom{5}{1} p^1 q^4 = 5p^1 q^4$

$P(X = 2) = \binom{5}{2} p^2 q^3 = 10p^2 q^3$

$P(X = 3) = \binom{5}{3} p^3 q^2 = 10p^3 q^2$ Remember that the powers of p and q add up to 5 each time.

$P(X = 4) = \binom{5}{4} p^4 q^1 = 5p^4 q^1$

$P(X = 5) = \binom{5}{5} p^5 q^0 = p^5$

The terms giving the **binomial probabilities** are the terms in the **binomial expansion** of $(q + p)^5$, where

$$(q + p)^5 = q^5 + 5p^1 q^4 + 10p^2 q^3 + 10p^3 q^2 + 5p^4 q^1 + p^5$$

$$ P(X = 0) \quad P(X = 1) \quad P(X = 2) \quad P(X = 3) \quad P(X = 4) \quad P(X = 5)$$

Check the sum of the probabilities:

$$\sum_{\text{all } x} P(X = x) = (q + p)^5 = 1^5 = 1$$ $q + p = 1$

Deciding whether a binomial distribution is appropriate

If you are given a practical situation and have to decide whether the random variable X follows a binomial distribution, check that **all** the conditions for a binomial distribution are satisfied **in the context of the question**. If any are not satisfied, then X does not follow a binomial distribution.

Example 5.8

In a bag there are 8 green counters and 7 red counters. Emma and Jack take part in an experiment in which 6 counters are to be selected at random from the bag. Emma is told to put the counter back into the bag after each trial and Jack is told to put the counter into his pocket after each trial.

The random variable X is the number of green counters selected. Explain why X follows a binomial distribution in Emma's experiment but not in Jack's experiment.

In both experiments the process is carried out 6 times, so the number of trials is fixed.

Also each trial results in a successful outcome (the counter is green) or an unsuccessful outcome (the counter is red).

Emma's experiment

The counter is put back into the bag after each selection. The outcome of a trial does not depend on the outcomes of preceding trials, so the trials are independent. Also, since $\text{P(green)} = \frac{8}{15}$ for each trial, the probability of success is constant. So X follows a binomial distribution in Emma's experiment.

> This is known as sampling **with** replacement.

Jack's experiment

The counter is not put back into the bag after each selection. The outcome of a trial depends on the outcomes of preceding trials, so the trials are not independent. X does not follow a binomial distribution.

> This is known as sampling **without** replacement.

Example 5.9

Kwasi is playing a board game. He has to throw a six on the die in order to start. The random variable X is the number of times Kwasi throws the die until he throws a six. Explain why X does not follow a binomial distribution.

Kwasi might get a six on his first throw, but he might need to throw the die many times to get a six.

Since there is no limit to the number of trials required, the number of trials is not fixed.

So X does not follow a binomial distribution.

---

*If you know only that a trial is repeated a fixed number of times, what assumptions **related to the context** do you need to make in order to use a binomial distribution?*

Example 5.10

Fred takes 12 shots at a goal.

State any assumptions necessary for the number of goals he scores to follow a binomial distribution.

The following assumptions must be made:

- Scoring a goal in one shot is independent of scoring a goal in all other shots.
- The probability of scoring a goal on any shot is the same for all 12 shots, that is, Fred's ability to score a goal does not improve with practice.

Exercise 5b

1 If $Y \sim B\left(6, \frac{1}{4}\right)$, find
   (i) $P(Y = 2)$,   (ii) $P(Y \geqslant 2)$.

2 If $M \sim B(9, 0.73)$, find
   (i) $P(M < 7)$,   (ii) $P(M > 6)$.

3 In a survey it is found that 65% of shoppers choose Soapy Suds when buying washing powder. A random sample of 10 shoppers buying washing powder is taken. Find the probability that no more than 8 shoppers in the sample choose Soapy Suds.

4 A 5-sided spinner is equally likely to stop on any of the numbers 1, 2, 3, 4 or 5. Kate spins it 10 times. Find the probability that the spinner stops on
   (i) an even number on exactly 7 spins,
   (ii) an odd number on more than 7 spins.

5 (i) A fair cubical die is thrown 12 times. Find the probability of throwing more than 2 sixes.
   (ii) Two fair cubical dice are thrown. Find the probability of throwing a double six (a six on each die).
   (iii) Two fair cubical dice are thrown 12 times. Show that the probability of throwing at least 3 double sixes is approximately 0.4%.

6 An experiment consists of taking shots at a target and counting the number of hits. The probability of hitting the target with a single shot is 0.8.

   Find the probability that in 13 consecutive attempts the target is hit at most 11 times.

7 If $X \sim B(4, p)$ and $P(X > 0) = 0.9744$, find
   (i) p (ii) $P(X = 2)$.

8 A coin is biased so that it is twice as likely to show heads as tails.
   (i) What is the probability that the coin will show heads when it is tossed?
   (ii) Andy tosses the coin n times. Find the least value of n for which the probability that the coin shows heads each time is less than 0.01.

9 Charlie finds that when she takes a cutting from a particular plant, the probability that it roots successfully is $\frac{1}{4}$.
   (i) She takes 9 cuttings. Find the probability that at least one cutting roots successfully.
   (ii) Charlie takes n cuttings. Find the smallest value of n if there is to be a probability of at least 0.99 that at least one cutting roots successfully.

10 On a production line making light bulbs, the probability of any light bulb being faulty is 0.05.

   The light bulbs are packed into boxes of 10.
   (i) A pack is selected at random. Show that the probability that the pack contains fewer than 2 faulty light bulbs is 0.9139, correct to 4 significant figures.

   Six packs of light bulbs are selected at random.
   (ii) Find the probability that exactly five of these packs contain fewer than 2 faulty light bulbs.

EXPECTATION AND VARIANCE OF THE BINOMIAL DISTRIBUTION

If $X \sim B(n, p)$, then

Expectation (mean)

$$E(X) = \mu = np$$

Variance

$$Var(X) = \sigma^2 = npq, \text{ where } q = 1 - p$$

Note that standard deviation $= \sigma = \sqrt{Var(X)} = \sqrt{npq}$

These results can be quoted and should be learned.

You will not need to know the proofs of these results.

In the following example the results are verified in a specific case.

Example 5.11

The random variable X has a binomial distribution with $n = 4$ and $p = 0.8$.

(i) Complete this table showing the probability distribution of X.

x	0	1	2	3	4
P(X = x)	0.0016		0.1536	0.4096	

(ii) Use the values in the table to calculate $E(X)$ and $Var(X)$.

(iii) Calculate np and npq, where $q = 1 - p$, and confirm that $E(X) = np$ and $Var(X) = npq$.

(iv) Find the standard deviation of X.

(i) $X \sim B(4, 0.8)$ with $n = 4$, $p = 0.8$, $q = 1 - 0.8 = 0.2$

$$P(X = 1) = \binom{4}{1} \times 0.8^1 \times 0.2^3 = 0.0256$$

$$P(X = 4) = 0.8^4 = 0.4096$$

The completed probability distribution table is shown below.

x	0	1	2	3	4
P(X = x)	0.0016	**0.0256**	0.1536	0.4096	**0.4096**

(ii) $E(X) = \mu = \sum xp$

See page **147**

$$= 0 \times 0.0016 + 1 \times 0.0256 + 2 \times 0.1536 + 3 \times 0.4096 + 4 \times 0.4096$$

$$= 3.2$$

$$Var(X) = \sum x^2 p - \mu^2$$

$$= 0^2 \times 0.0016 + 1^2 \times 0.0256 + 2^2 \times 0.1536$$
$$+ 3^2 \times 0.4096 + 4^2 \times 0.4096 - 3.2^2$$

Remember to subtract μ^2.

$$= 10.88 - 3.2^2$$

$$= 0.64$$

(iii) $np = 4 \times 0.8 = 3.2 = E(X)$ from part (ii)

$npq = 4 \times 0.8 \times 0.2 = 0.64 = Var(X)$ from part (ii)

This confirms that $E(X) = np$ and $Var(X) = npq$.

(iv) Standard deviation $= \sqrt{npq} = \sqrt{0.64} = 0.8$

Example 5.12

The mean number of defective batteries in packs of 20 is 1.6. Use a binomial distribution to calculate the probability that a randomly chosen pack of 20 will have more than 2 defective batteries.

Cambridge Paper 61 Q1 N09

Let X be the number of defective batteries in a pack of 20.

$X \sim \text{B}(n, p)$ with $n = 20$, p is unknown.

Mean $= np = 20 \times p$

But you are told that the mean number of defective batteries is 1.6,

so $20 \times p = 1.6$

$$p = \frac{1.6}{20} = 0.08$$

Hence $X \sim \text{B}(20, 0.08)$ with $n = 20$, $p = 0.08$, $q = 0.92$

$\text{P}(X > 2)$

$= 1 - (\text{P}(X = 0) + \text{P}(X = 1) + \text{P}(X = 2))$

$= 1 - \left(0.92^{20} + \binom{20}{1} \times 0.08^1 \times 0.92^{19} + \binom{20}{2} \times 0.08^2 \times 0.92^{18} \right)$

$= 1 - 0.7879....$

$= 0.2120...$

$= 0.212$ (3 s.f.)

Example 5.13

The random variable X is distributed $\text{B}(n, p)$ with mean 5 and standard deviation 2.
Find the values of n and p.

$np = \mu = 5$ (1)

$npq = \sigma^2 = 2^2 = 4$ (2)   Remember to square the standard deviation.

Substitute for np from (1) into (2):

$5q = 4$

$q = 0.8$

$p = 1 - q = 0.2$

Substitute for p in (1)

$n \times 0.2 = 5$

$$n = \frac{5}{0.2} = 25$$

So $n = 25$ and $p = 0.2$.

Example 5.14

Lilia travels to work by bus or by car. The probability that she travels by bus on any day is 0.7. If she travels by bus, there is a probability of 0.1 that she is late for work. If she travels by car there is a probability of 0.2 that she is late for work.

(i) Find the probability that she is late for work on a particular day.

(ii) Find the expected number of days she is late for work in 20 working days.

(iii) Find the variance of the number of days she travels by car in 10 working days.

(i) *Show the information on a tree diagram.*

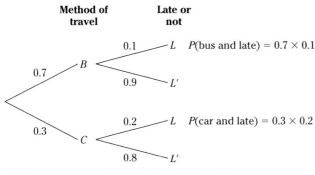

$$P(\text{Lilia is late}) = (0.7 \times 0.1) + (0.3 \times 0.2)$$
$$= 0.13$$

(ii) Let X be the number of days Lilia is late in 20 days.
$X \sim B(20, 0.13)$ with $n = 20$, $p = 0.13$ from part (i)

$E(X) = np = 20 \times 0.13 = 2.6$

The expected number of days she is late is 2.6.

(iii) $P(\text{Lilia travels by car}) = 1 - 0.7 = 0.3$

Let Y be the number of days Lilia travels by car in 10 days.
$Y \sim B(10, 0.3)$ with $n = 10$, $p = 0.3$, $q = 0.7$

Variance of $Y = npq = 10 \times 0.3 \times 0.7 = 2.1$

DIAGRAMMATIC REPRESENTATION OF THE BINOMIAL DISTRIBUTION

It can be useful to have an idea of the shape of the binomial distribution.

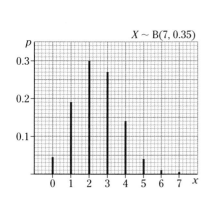

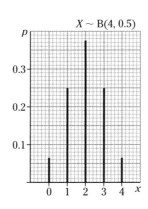

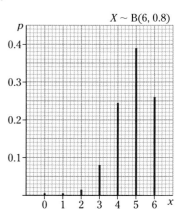

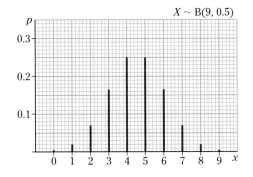

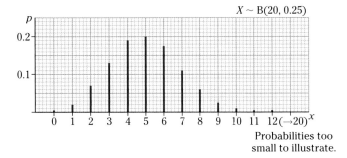

Notice that the binomial distribution is symmetrical when $p = 0.5$.

Probabilities too small to illustrate.

Example 5.15

The random variable X has distribution B(20, 0.25).

This distribution is illustrated in the last diagram above

(i) Find the mean and the standard deviation of X.

(ii) Find the percentage of the distribution that lies within one standard deviation of the mean.

$X \sim$ B(20, 0.25) with $n = 20$, $p = 0.25$, $q = 1 - p = 0.75$

(i) $\mu = np = 20 \times 0.25 = 5$

$\sigma^2 = \text{Var}(X) = npq = 20 \times 0.25 \times 0.75 = 3.75$

$\sigma = \sqrt{3.75} = 1.936... = 1.94$ (3 s.f.)

(ii) P(X lies within one standard deviation of the mean)

$= \text{P}(\mu - \sigma < X < \mu + \sigma)$

$= \text{P}(5 - 1.936... < X < 5 + 1.936...)$

$= \text{P}(3.063... < X < 6.936...)$

$= \text{P}(X = 4) + \text{P}(X = 5) + \text{P}(X = 6)$

$= \binom{20}{4} \times 0.25^4 \times 0.75^{16} + \binom{20}{5} \times 0.25^5 \times 0.75^{15} + \binom{20}{6} \times 0.25^6 \times 0.75^{14}$

$= 0.5606...$

$= 56.1\%$ (3 s.f.)

So 56.1% of the distribution lies within one standard deviation of the mean.

THE MODE OF THE BINOMIAL DISTRIBUTION

The **mode** is the value of X that is most likely to occur, i.e. the value with the **highest probability**. In a vertical line diagram, the mode is the value represented by the highest line.

Consider the distributions shown in the diagrams above and locate the mode. You will notice, however, that $X \sim$ B(9, 0.5) has two modes, i.e. it is bi-modal. In fact this is the case for any binomial distribution where n is odd and $p = 0.5$.

In general,
– when n is odd and $p = 0.5$, the binomial distribution has two modes
– for all other values of n and p, the binomial distribution has one mode.

The mode can be found by calculating all the probabilities and finding the value of X with the highest probability. This is however very tedious; it is usually only necessary to consider the probabilities of values of X close to the mean of the distribution.

Example 5.16

The probability that a student at a particular college is awarded a distinction is 0.05. The number of students awarded a distinction in a randomly chosen group of 50 students from the college is denoted by X.

(i) Find the mean of X.

(ii) Find the most likely value of X.

X is the number of students in 50 awarded a distinction,
so $X \sim B(50, 0.05)$.

(i) mean $= E(X) = np = 50 \times 0.05 = 2.5$

(ii) *To find the most likely value, you have to find the value of X with the highest probability. Consider values close to the mean.*

$$P(X = 1) = \binom{50}{1} \times 0.05^1 \times 0.95^{49} = 0.202\ldots$$

$$P(X = 2) = \binom{50}{2} \times 0.05^2 \times 0.95^{48} = 0.261\ldots$$

$$P(X = 3) = \binom{50}{3} \times 0.05^3 \times 0.95^{47} = 0.219\ldots$$

From the list you can see that the value of X with the highest probability is 2.

The most likely value of X is 2.

Exercise 5c

1 The random variable X has distribution B(14, 0.36). Find

  (i) the mean,

  (ii) the variance,

  (iii) the standard deviation.

2 The random variable Y has distribution B(20, 0.4). Find the probability that Y is equal to the mean of Y.

3 In a large consignment of apples, 15% are rejected for being too small. A random sample of 20 apples is taken from the consignment.

  (i) Find the expected number of rejected apples in the sample.

  (ii) Explain why the most likely number of rejected apples in the sample is 3.

4 The probability that an item produced by a machine is satisfactory is 0.92.

  (i) Find the expected number of satisfactory items in a random sample of 25 items produced by the machine.

  (ii) Find the standard deviation of the number of **unsatisfactory** items in a random sample of 50 produced by the machine.

5 The random variable Y has distribution B(n, 0.3) and $E(Y) = 2.4$.

  (i) Find n.

  (ii) Find $P(Y = 5)$.

  (iii) Find the standard deviation of Y.

6 X is the number of tails when an unbiased coin is tossed 10 times. The mean of X is μ and the standard deviation of X is σ.

  (i) Find μ and σ.

  (ii) Find $P(X < \mu - 2\sigma)$.

  (iii) Find $P(X > \mu + 2\sigma)$.

  (iv) Find the probability that X is **more than** two standard deviations away from the mean.

7 In a multiple choice test, for each question students have to choose the correct answer from a choice of four answers. There are 20 questions in the test. Jack decides not to read any of the questions, but to select an answer at random each time.

(i) What is the probability that Jack answers any one question correctly?

(ii) Find the mean and standard deviation of the number of correct answers that Jack gets.

(iii) Find the probability that the number of correct answers that Jack gets is within one standard deviation of the mean.

8 In a bag there are 6 red counters, 8 yellow counters and 6 green counters.

Ronami selects a counter at random from the bag, notes its colour and then puts it back into the bag. She does this four times in all.

Find

(i) the probability that she selects 4 red counters,

(ii) the expected number of yellow counters she selects,

(iii) the variance of the number of green counters she selects.

9 The probability that a person chosen at random wears glasses is p. A random sample of n people is chosen and the number of people in the sample who wear glasses is denoted by X. It is given that $E(X) = 2.4$ and $\text{Var}(X) = 1.68$.

(i) Find the value of p.

(ii) Find the value of n.

(iii) Find the probability that exactly 6 people in the sample wear glasses.

10 It is given that $X \sim B(n, p)$. The mean of X is 3.6 and the variance of X is 2.16. Find

(i) n and p,   (ii) $P(X \leqslant 2)$.

11 A calculator generates the digits 0, 1, 2, 3, 4, 5, 6, 7, 8 and 9 randomly so that each digit has an equal chance of occurring.

(i) What is the probability that a zero is generated?

(ii) Six digits are generated. Find the probability that there is exactly one zero.

(iii) Twenty digits are generated. Find the expected number of digits that are multiples of 3.

(iv) One hundred digits are generated. Find the variance of the number of odd digits.

12 The discrete random variable X is such that $X \sim B(n, p)$ with $n = 3$ and $p = 0.4$.

(i) Complete the probability distribution table for X.

x	0	1	2	3
$P(X = x)$		0.432		0.064

(ii) Use the values in the table to calculate $E(X)$ and $\text{Var}(X)$.

(iii) Calculate np and npq, where $q = 1 - p$, and confirm that $E(X) = np$ and $\text{Var}(X) = npq$.

Summary

If X is the number of successful outcomes in n independent trials and p is the probability of a successful outcome, then $X \sim B(n, p)$.

$$P(X = r) = \binom{n}{r}p^r q^{n-r} \quad \text{where } q = 1 - p, \quad \text{for } r = 0, 1, 2, \ldots, n$$

Number of ways to choose r from n Probability of r successes   Probability of $(n - r)$ failures

$$\binom{n}{r} = {}_nC_r = \frac{n!}{r!(n - r)!}$$

Expectation $E(X) = \mu = np$

Variance $\text{Var}(X) = \sigma^2 = npq$

Standard deviation $\sigma = \sqrt{npq}$

The most likely number of successes (the mode) is the value of X with the highest probability.

Mixed Exercise 5

There are further questions relating to the binomial distribution in Chapter 6.

1 If $X \sim B(11, 0.65)$, find

    (i) $P(X = 6)$,

    (ii) $P(X < 3)$,

    (iii) $P(X \geqslant 9)$.

2 The probability that a component produced by a particular machine is defective is 0.05. A random sample of 12 components is selected from the production line of the machine. Find the probability that the number of defective components in the sample is

    (i) exactly 3,   (ii) at least 2.

3 On average, 1 in 8 people living in a particular country were not born in that country. In a randomly selected group of 20 people living in the country, find the probability that

    (i) exactly 5 were not born in the country,

    (ii) at least 3 were not born in the country,

    (iii) more than 18 were born in the country.

4 The random variable X has a binomial distribution with $n = 30$ and $p = 0.5$. Find

    (i) $P(12 < X < 15)$,

    (ii) $P(X = E(X))$.

5 The random variable X has the distribution $B(9, 0.45)$. Find

    (i) $E(X)$,

    (ii) the standard deviation of X.

6 The random variable X has a binomial distribution with $n = 16$ and $p = 0.15$. Find

    (i) $E(X)$,

    (ii) the probability that X is greater than $E(X)$,

    (iii) $Var(X)$.

7 In the holiday period, the probability that Peter plays tennis on any particular day is $\frac{2}{3}$.

    (i) Find the probability that Peter plays tennis on exactly 5 days in a holiday period of 14 days.

    (ii) Find the mean number of days on which Peter plays tennis in a holiday period of 21 days.

    (iii) If the standard deviation of the number of days Peter plays tennis is $\frac{4}{3}$, how many days are there in the holiday period?

8 The table shows the probability distribution of the discrete random variable X.

x	0	1	2	3
$P(X = x)$	0.4	0.3	0.2	0.1

    (i) Find the probability that a random observation from X is odd.

    (ii) Ten random observations of X are made. Find the mean number of observations that are odd numbers.

9 Balloons are packaged in party bags. Each bag contains 20 balloons. The colours of the balloons in a party bag are shown in the table.

Colour	Red	Blue	Green	Yellow
Frequency	8	5	4	3

Serene buys 10 party bags of balloons and selects a balloon at random from each bag.

    (i) Find the probability that she selects at least 2 green balloons.

    (ii) Find the mean number of blue balloons that she selects.

    (iii) Find the variance of the number of yellow balloons she selects.

10 The probability that Jenny receives at least one telephone call on any day is 0.8. The number of telephone calls she receives is independent from day to day.

    (i) Calculate the probability that, during a particular fortnight, Jenny receives at least one telephone call on exactly 9 days.

    (ii) Calculate the mean number of days in April on which she does not receive any telephone calls.

11 Crocus and tulip bulbs are sold in mixed packs of 36 bulbs. On average, a pack contains three times as many crocus bulbs as tulip bulbs.

(i) A pack is selected at random. Find the probability that two-thirds of the bulbs in the pack are crocus bulbs.

(ii) Find the variance of the number of tulip bulbs in a pack.

12 A fair cubical die has one face numbered 1, two faces numbered 2 and three faces numbered 3.

Find the probability of obtaining fewer than 6 odd numbers in 7 throws of the die.

6 The normal distribution

In this chapter you will learn

- about the normal variable X and its distribution
- about the standard normal variable Z
- how to use the normal distribution table to find probabilities
- how to standardise any normal variable X
- how use the normal distribution table in reverse
- how to use the normal distribution as an approximation to the binomial distribution
- how to use continuity corrections

CONTINUOUS RANDOM VARIABLES

A **continuous random variable** cannot take precise values but can only be given within a specified interval. It usually arises from **measuring** a characteristic such as time, mass or length. Examples are the time between cars passing a checkpoint, the error made by scales when weighing potatoes, the length of leaves on a particular type of bush.

A continuous random variable X is defined by its probability density function f, together with the values between which the function is valid, for example $a < x < b$.

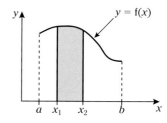

Areas beneath the graph of $y = f(x)$ represent probabilities, so $P(x_1 \leqslant X \leqslant x_2)$ is equal to the area beneath the graph between $x = x_1$ and $x = x_2$.

The total area beneath the graph of $y = f(x)$ is 1.

In module S1 the distribution of a special type of continuous random variable is studied: the normal distribution. Continuous random variables in general are studied in module S2.

THE NORMAL DISTRIBUTION

One of the most important continuous random variables in statistics is the **normal variable.** Its distribution, known as the **normal distribution**, is a good model for many situations, especially in the natural sciences. Also, in certain circumstances it can be used as an approximation to other distributions such as the binomial distribution.

In a normal distribution there are relatively few very small or very large values, with the bulk of the distribution being concentrated around the middle value.

Here are some variables that might be modelled by a normal distribution:

- The height of adult males in a particular country
- The time taken by students to complete a task
- The volume of coffee dispensed into a cup by a particular coffee machine
- The mass of babies at birth

A normal distribution is specified completely by stating its mean μ and variance σ^2.

Note that the standard deviation is σ.

Remember that the standard deviation is the square root of the variance.

If X follows a normal distribution, you can write

$$X \sim N(\mu, \sigma^2).$$

mean variance

This is read as

'X has a normal distribution with mean μ and variance σ^2.'

If X follows a normal distribution, the graph of $y = f(x)$ is a bell-shaped curve which is symmetrical about the mean μ.

The total area under the normal curve is 1.

In theory the possible values of x are from $-\infty$ to $+\infty$,

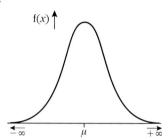

However, in practice:

- approximately 68% (just over two-thirds) of the distribution lies within 1 standard deviation of the mean, i.e. between $\mu - \sigma$ and $\mu + \sigma$

- approximately 95% of the distribution lies within 2 standard deviations of the mean, i.e. between $\mu - 2\sigma$ and $\mu + 2\sigma$

- approximately 99.7% of the distribution (almost all) lies within 3 standard deviations of the mean, i.e. between $\mu - 3\sigma$ and $\mu + 3\sigma$.

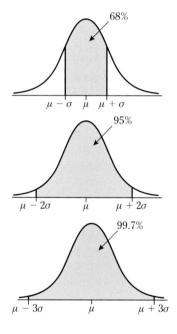

Note that
- changes in the mean μ alter the position of the curve along the x-axis
- changes in the standard deviation σ alter the spread of the curve about the mean.

Here are some normal curves drawn to the same scale.

(i) $X \sim N(0, 1^2)$
  $\mu = 0, \sigma = 1$

(ii) $X \sim N\left(4, \left(\frac{1}{2}\right)^2\right)$
  $\mu = 4, \sigma = \frac{1}{2}$

(iii) $X \sim N(50, 2^2)$
  $\mu = 50, \sigma = 2$

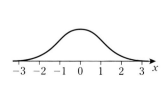

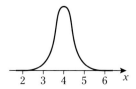

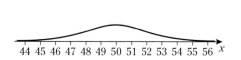

Finding probabilities

Probabilities are given by **areas** under the normal curve.

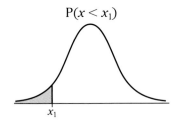

$P(x < x_1)$

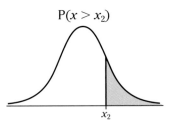

$P(x > x_2)$

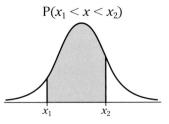

$P(x_1 < x < x_2)$

Note that $P(x = x) = 0$ for all values of x. This means that $P(x < x_1)$ is the same as $P(x \leqslant x_1)$, $P(x > x_2)$ is the same as $P(x \geqslant x_2)$ and $P(x_1 < x < x_2)$ is the same as $P(x_1 \leqslant x \leqslant x_2)$.

A table is used to find areas, and hence probabilities. It is written for one particular normal variable called the standard normal variable. However, it can be used for **any** normal variable by a process called standardising, described later on page **184**.

The standard normal variable Z

The **standard normal variable** is given the special symbol Z, rather than x. It has a mean of 0 and a variance of 1, so $Z \sim N(0, 1)$.

Notice that the standard deviation of Z is also 1.

The normal distribution table

If Z has a normal distribution with mean 0 and variance 1 then, for each value of z, the table gives the value of $\Phi(z)$, where

$$\Phi(z) = P(Z \leqslant z) = P(Z < z)$$

Φ is the Greek letter 'phi'.   Remember that there is no distinction between $P(Z \leqslant z)$ and $P(Z < z)$.

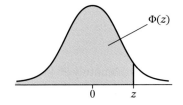

$\Phi(z)$

The **normal distribution table**, which is provided in the examination, is printed on page **213**. An extract is shown below. The highlighted values relate to the illustrations that follow.

	z	0	1	2	3	4	5	6	7	8	9	1	2	3	4	5	6	7	8	9
												\multicolumn{9}{c}{ADD}								
	0.0	0.5000	0.5040	0.5080	0.5120	0.5160	0.5199	0.5239	0.5279	0.5319	0.5359	4	8	12	16	20	24	28	32	36
(i)	0.1	0.5398	0.5438	0.5478	0.5517	0.5557	0.5596	0.5636	0.5675	0.5714	0.5753	4	8	12	16	20	24	28	32	36
	0.2	0.5793	0.5832	0.5871	0.5910	0.5948	0.5987	0.6026	0.6064	0.6103	0.6141	4	8	12	15	19	23	27	31	35
(ii)	0.3	0.6179	0.6217	0.6255	0.6293	0.6331	0.6368	0.6404	0.6443	0.6480	0.6517	4	7	11	15	19	22	26	30	34

To use the table, follow these instructions.

(i) $P(Z < 0.16)$
   $= \Phi(0.16)$
   $= 0.5636$

Find row 0.1.
Go across to column 6.
This gives **0.5636**.

(ii) $P(Z < 0.345)$
   $= \Phi(0.345)$
   $= 0.6350$

Find row 0.3 and go across to column 4. This gives 0.6331.

Continue along the row to column 5 in the section on the far right. This gives 19.

ADD 19 to the **digits** 6331.

The required value is then **0.6350**.

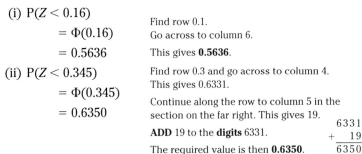

$$\begin{array}{r} 6331 \\ + \quad 19 \\ \hline 6350 \end{array}$$

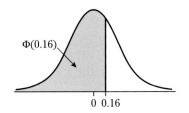

$\Phi(0.16)$

0 0.16

Note

If the z-value has more than 3 decimal places you need to round it to 3 decimal places first. For example if $z = 1.3469$, round it to 1.347 and find $\Phi(1.347)$.

Values of $\Phi(z)$, and hence probabilities, are given to 4 decimal places in the main body of the table.

To find P($Z > z$) use the fact that the total area under the curve is 1.

So   P($Z > z$) = 1 − Φ(z)

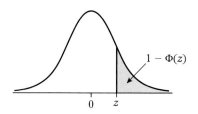

Example 6.1

If $Z \sim N(0, 1)$ find

(i) P($Z < 0.85$)

(ii) P($Z > 0.85$).

(i)

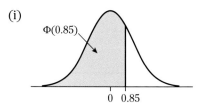

P($Z < 0.85$) = Φ(0.85)
            = 0.8023

(ii)

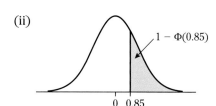

P($Z > 0.85$) = 1 − Φ(0.85)
            = 1 − 0.8023
            = 0.1977

You will notice that the table starts at $z = 0$ and gives values for positive values of z. When z is negative you need to use the fact that the normal curve is **symmetrical**.

In the following illustrations, a is positive.

By symmetry,

P($Z < -a$) = P($Z > a$)

So,

P($Z < -a$) = Φ($-a$) = 1 − Φ(a)

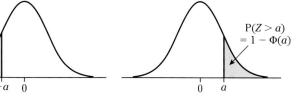

Also by symmetry,

P($Z > -a$) = P($Z < a$)

So,

P($Z > -a$) = Φ(a)

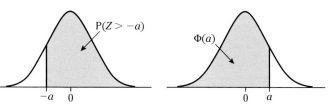

Example 6.2

If $Z \sim N(0, 1)$ find

(i) P($Z < 1.377$)

(ii) P($Z > -1.377$)

(iii) P($Z > 1.377$)

(iv) P($Z < -1.377$).

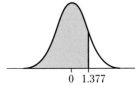

(i)

$P(Z < 1.377) = \Phi(1.377)$
$= 0.9158$

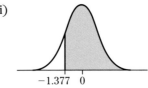

(ii)

$P(Z > -1.377) = \Phi(1.377)$
$= 0.9158$

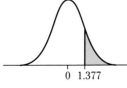

(iii)

$P(Z > 1.377) = 1 - \Phi(1.377)$
$= 1 - 0.9158$
$= 0.0842$

(iv)

$P(Z < -1.377) = 1 - \Phi(1.377)$
$= 1 - 0.9158$
$= 0.0842$

Important results

These are worth learning.

In the following illustrations, a and b are positive and $a < b$.

Result 1

$$P(a < Z < b) = P(Z < b) - P(Z < a)$$
$$= \Phi(b) - \Phi(a)$$

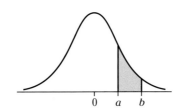

The area could be described as a 'right-hand sandwich'.

So $P(a < Z < b) = \Phi(b) - \Phi(a)$

Result 2

$$P(-b < Z < -a) = P(Z < -a) - P(Z < -b)$$
$$= \Phi(-a) - \Phi(-b)$$
$$= 1 - \Phi(a) - (1 - \Phi(b))$$
$$= 1 - \Phi(a) - 1 + \Phi(b)$$
$$= \Phi(b) - \Phi(a)$$

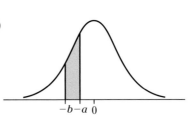

A 'left-hand sandwich'

So $P(-b < Z < -a) = \Phi(b) - \Phi(a)$

Example 6.3

If $Z \sim N(0, 1)$, find
  (i) $P(0.5 < Z < 2)$
  (ii) $P(-1.4 < Z < -0.6)$

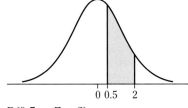

(i)

P(0.5 < Z < 2)
$$= \Phi(2) - \Phi(0.5)$$
$$= 0.9772 - 0.6915$$
$$= 0.2857$$

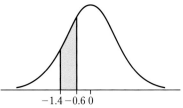

(ii)

P(−1.4 < Z < −0.6)
$$= \Phi(1.4) - \Phi(0.6)$$
$$= 0.9192 - 0.7257$$
$$= 0.1935$$

Result 3

$$P(-a < Z < b) = P(Z < b) - P(Z < -a)$$
$$= \Phi(b) - \Phi(-a)$$
$$= \Phi(b) - (1 - \Phi(a))$$

A 'middle sandwich'

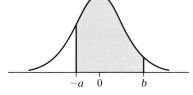

So $P(-a < Z < b) = \Phi(b) - (1 - \Phi(a))$

Result 4

A 'symmetrical middle sandwich'

$$P(-a < Z < a) = \Phi(a) - (1 - \Phi(a))$$ Expand the bracket
$$= \Phi(a) + \Phi(a) - 1$$
$$= 2\Phi(a) - 1$$

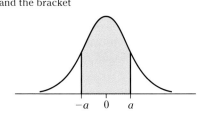

So $P(-a < Z < a) = 2\Phi(a) - 1$

Note that $P(-a < Z < a)$ may also be written $P(|Z| < a)$.

This is a special case of Result 3, obtained by expanding the brackets.

Example 6.4

If $Z \sim N(0, 1)$
 (i) find $P(-2.696 < Z < 1.865)$.
 (ii) Show that the central 95% of the distribution lies between $z = -1.96$ and $z = 1.96$.

 (i) P(−2.696 < Z < 1.865)
$$= \Phi(1.865) - \Phi(-2.696)$$
$$= \Phi(1.865) - (1 - \Phi(2.696))$$
$$= 0.9690 - (1 - 0.9965)$$
$$= 0.9655$$

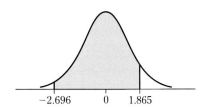

(ii) $P(-1.96 < Z < 1.96)$

$\quad = \Phi(1.96) - \Phi(-1.96)$

$\quad = \Phi(1.96) - (1 - \Phi(1.96))$

$\quad = 0.975 - (1 - 0.975)$

$\quad = 0.95$

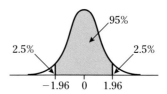

Alternatively, you could expand the bracket first to get

$\quad P(-1.96 < Z < 1.96)$

$\quad\quad = 2\Phi(1.96) - 1$

$\quad\quad = 2 \times 0.975 - 1$ It is useful to be able
to quote Result 4.

$\quad\quad = 0.95$

So, the central 95% of the distribution lies between $z = -1.96$ and $z = 1.96$ This is sometimes
i.e. between $z = \pm 1.96$.   written z = ±1.96

Example 6.5

Given that $P(Z > z_1) = 0.65$ and $P(Z > z_2) = 0.25$, find $P(z_1 < Z < z_2)$.

*It is helpful to locate the positions of z_1 and z_2 on a sketch.
Think carefully about whether they are positive (in the
upper tail to the right of 0) or negative (in the lower tail to
the left of 0).*

If $P(Z > z_1) = 0.65$, more than half
the area is to the right of z_1, so z_1
must be to the left of the middle, i.e.
in the lower tail.

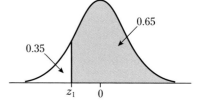

Note that $\Phi(z_1) = 0.35$.

If $P(Z > z_2) = 0.25$, less than half the area
is to the right of z_2, so z_2 must be to the
right of the middle, i.e. in the upper tail.

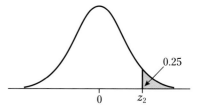

Now put the information on one diagram:
$P(z_1 < Z < z_2)$

$\quad = 1 - (0.25 + 0.35)$

$\quad = 0.4$

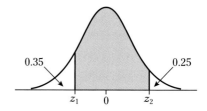

Exercise 6a

In all the following questions $Z \sim N(0, 1)$.

Always **draw sketches** to illustrate the probabilities and also check that your answer is sensible. For example, if you get a negative answer for a probability it is definitely wrong. It is also wrong if you get a probability greater than 1.

1 Find

  (i) $P(Z < 0.874)$ (ii) $P(Z > -0.874)$

  (iii) $P(Z > 0.874)$ (iv) $P(Z < -0.874)$

2 Find

  (i) $P(Z > 1.8)$ (ii) $P(Z < -0.65)$

  (iii) $P(Z \geqslant -2.46)$ (iv) $P(Z \leqslant 1.36)$

  (v) $P(Z > 2.58)$ (vi) $P(Z > -2.37)$

  (vii) $P(Z < 1.86)$ (viii) $P(Z \leqslant -0.725)$

  (ix) $P(Z > 1.863)$ (x) $P(Z < 1.63)$

  (xi) $P(Z > -2.061)$ (xii) $P(Z < -2.875)$

3 Find

  (i) $P(Z \geqslant 1.645)$ (ii) $P(Z < -1.645)$

  (iii) $P(Z > 1.282)$ (iv) $P(Z > 1.96)$

  (v) $P(Z > -2.575)$ (vi) $P(Z \geqslant 2.326)$

  (vii) $P(Z \leqslant -2.808)$ (viii) $P(Z < 1.96)$

4 Find the probabilities represented by the shaded areas in the diagrams.

  (i)

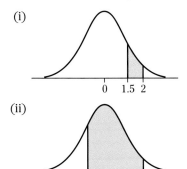

  (ii)

  (iii)

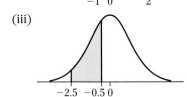

  (iv)

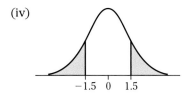

5 Find

  (i) $P(0.829 < Z < 1.834)$

  (ii) $P(-2.56 < Z < 0.134)$

  (iii) $P(-1.762 \leqslant Z \leqslant -0.246)$

  (iv) $P(0 < Z < 1.73)$

  (v) $P(-2.05 < Z < 0)$

  (vi) $P(-2.08 < Z < 2.08)$

  (vii) $P(1.764 < Z < 2.567)$

  (viii) $P(-1.65 \leqslant Z < 1.725)$

  (ix) $P(-0.98 < Z < -0.16)$

  (x) $P(Z < -1.97 \text{ or } Z > 2.5)$

6 Find

  (i) $P(-1.78 < Z < 1.78)$

  (ii) $P(-1.645 < Z < 1.645)$

7 Complete this statement:

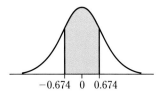

The central ......% of the distribution lies between -0.674 and 0.674.

8 It is given that $P(Z < a) = 0.3$ and $P(a < Z < b) = 0.6$.
  Find   (i) $P(Z < b)$ (ii) $P(Z > a)$

9 It is given that $P(Z < a) = 0.7$ and $P(Z > b) = 0.45$.
  Find   (i) $\Phi(b)$ (ii) $P(b < Z < a)$

10 It is given that $P(-a < Z < a) = 0.8$.
    Find   (i) $P(Z < a)$ (ii) $P(Z > a)$

STANDARDISING ANY NORMAL VARIABLE X

To use the tables for any normal variable X with mean μ and variance σ^2, the variable

$$X \sim N(\mu, \sigma^2)$$

is **standardised** (linearly scaled) to

μ is scaled to 0
σ^2 is scaled to 1

$$Z \sim N(0, 1)$$

This is illustrated below:

Figure 1:

This shows $y = f(x)$ where

$$X \sim N(\mu, \sigma^2)$$

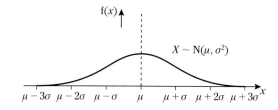

Figure 2:

Translate the curve by subtracting μ
so that the mean is 0.
You now have

$$X - \mu \sim N(0, \sigma^2).$$

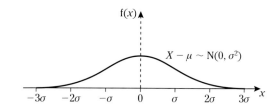

Figure 3:

Divide by the standard deviation σ.
This reduces the horizontal scale by a factor of σ.
The standard deviation, and hence the variance, is now 1.
You now have

$$Z = \frac{X - \mu}{\sigma} \sim N(0, 1)$$

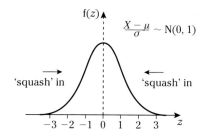

In general, the standardised value z tells you how many standard deviations the x-value is **above** the mean (z is **positive**) or **below** the mean (z is **negative**).

To standardise any normal variable $X \sim N(\mu, \sigma^2)$
- subtract the mean μ
- then divide by the standard deviation σ

to obtain

$$Z = \frac{X - \mu}{\sigma} \text{ where } Z \sim N(0, 1)$$

Example 6.6

Metal strips are produced by a machine. The lengths of strips follow a normal distribution with mean 150 cm and standard deviation 10 cm.

Find the probability that a randomly chosen strip from the production line has a length

(i)   less than 165 cm,

(ii) between 127.5 cm and 139.2 cm,

(iii) that deviates from the mean by more than 28 cm.

Let X be the length, in cm, of a metal strip.   *Define the variable in words and state its distribution.*

$X \sim N(150, 10^2)$ where $\mu = 150$ and $\sigma = 10$.

(i)   *To find the corresponding z-value, subtract the mean and divide by the standard deviation.*

$P(X < 165)$

$= P\left(Z < \dfrac{165 - 150}{10}\right)$

$= P(Z < 1.5)$

$= \Phi(1.5)$

$= 0.9332$

$= 0.933$ (3 s.f.)

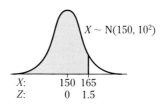

Note that although the variables X and Z have different means and spreads, it is convenient to show the corresponding values for them on **one** sketch.

Unless instructed otherwise, give your **final** answer to 3 significant figures.

(ii) $P(127.5 < X < 139.2)$

$= P\left(\dfrac{127.5 - 150}{10} < Z < \dfrac{139.2 - 150}{10}\right)$

$= P(-2.25 < Z < -1.08)$

$= \Phi(2.25) - \Phi(1.08)$

$= 0.9878 - 0.8599$ ← Result 2 for a 'left-hand sandwich' page **180**.

$= 0.1279$

$= 0.128$ (3 s.f.)

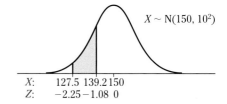

(iii) The length 28 cm below the mean is 122 cm and the length 28 cm above the mean is 178 cm.

So, P(the length deviates from the mean by more than 28 cm)

$= P(X < 122 \text{ or } X > 178)$

First find P(122 < X < 178) and then subtract this probability from 1 (the total probability).

$P(122 < X < 178)$

$= P\left(\dfrac{122 - 150}{10} < Z < \dfrac{178 - 150}{10}\right)$

$= P(-2.8 < Z < 2.8)$

$= 2\Phi(2.8) - 1$ ← Result 4 for a 'symmetrical middle sandwich' page **181**.

$= 2 \times 0.9974 - 1$

$= 0.9948$

So P($X < 122$ or $X > 178$)

$= 1 - 0.9948$

$= 0.0052$

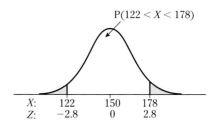

Notice that just over 0.5% of the metal strips deviate from the mean by more than 28 cm.

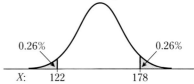

Example 6.7

Every day Bruno jogs around the park. The time he takes, in minutes, follows a normal distribution with mean 12 and variance 2.

(i) Find the probability that he takes longer than 14 minutes.

(ii) Find the probability that he takes less than 9 minutes.

(iii) Estimate the number of days during a year that he takes between 10 and 13 minutes.

X is the time, in minutes, that Bruno takes to jog around the park.

Define the variable in words and state its distribution.

$X \sim N(12, 2)$ where $\mu = 12$ and $\sigma^2 = 2$, so $\sigma = \sqrt{2}$.

Note: To avoid rounding errors, do not find a numerical value for $\sqrt{2}$, but input it directly on the calculator using $\sqrt{\ }$.

Always check whether you are given the variance or the standard deviation.

(i) $P(X > 14) = P\left(Z > \dfrac{14 - 12}{\sqrt{2}}\right)$

$= P(Z > 1.414)$

$= 1 - \Phi(1.414)$

$= 1 - 0.9213$

$= 0.0787$

Standardise by subtracting the mean and dividing by the standard deviation.

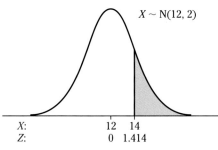

(ii) $P(X < 9) = P\left(Z < \dfrac{9 - 12}{\sqrt{2}}\right)$

$= P(Z < -2.121)$

$= 1 - \Phi(2.121)$

$= 1 - 0.9830$

$= 0.0170$

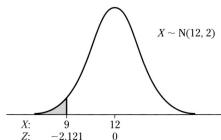

(iii) *First find the probability that he takes between 10 and 13 minutes.*

$P(10 < X < 13) = P\left(\dfrac{10 - 12}{\sqrt{2}} < Z < \dfrac{13 - 12}{\sqrt{2}}\right)$

$= P(-1.414 < Z < 0.707)$

$= \Phi(0.707) - (1 - \Phi(1.414))$

$= 0.7601 - (1 - 0.9213)$

$= 0.6814$

Result 3 for a 'middle sandwich' page **181**.

You have found the proportion of days (0.6814 or 68.14%) so to find the number of days in a *year* that Bruno takes between 10 and 13 minutes, multiply by *365*:

$0.6814 \times 365 = 248.711 \approx 249$

Bruno takes between 10 and 13 minutes on approximately 249 days in the year.

Example 6.8

A normal distribution has mean μ and standard deviation σ. If 800 observations are taken from the distribution, how many would you expect to be between $\mu - \sigma$ and $\mu + \sigma$?

Cambridge Paper 6 Q3(b) J07

$X \sim N(\mu, \sigma^2)$

$\mu - \sigma$ is 1 standard deviation below the mean, so $z = -1$.

$\mu + \sigma$ is 1 standard deviation above the mean, so $z = 1$.

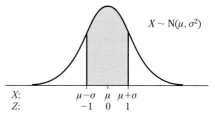

$X \sim N(\mu, \sigma^2)$

$$P(\mu - \sigma < X < \mu + \sigma)$$
$$= P(-1 < Z < 1)$$ Result 4 for a symmetrical middle sandwich, page **181**.
$$= 2\Phi(1) - 1$$
$$= 2 \times 0.8413 - 1$$
$$= 0.6826 = 68.26\%$$

Now 68.26% of 800 = 546.08 ≈ 546

See page **177**.

So you would expect approximately 546 observations to be between $\mu - \sigma$ and $\mu + \sigma$.

It is useful to remember the following:

$$P(\mu - \sigma < X < \mu + \sigma)$$
$$= P(-1 < Z < 1)$$
$$\approx 68\% \text{ (just over two-thirds)}$$

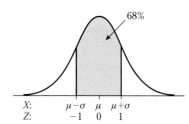

$$P(\mu - 2\sigma < X < \mu + 2\sigma)$$
$$= P(-2 < Z < 2)$$
$$\approx 95\%$$

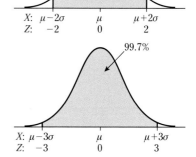

$$P(\mu - 3\sigma < X < \mu + 3\sigma)$$
$$= P(-3 < Z < 3)$$
$$\approx 99.7\% \text{ (almost all)}$$

Sketching a normal curve

Suppose you want to sketch the curve for $X \sim N(300, 25)$.

You know that the mean μ is 300 and the standard deviation σ is 5.

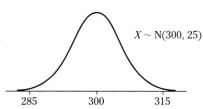

$X \sim N(300, 25)$

You also know that almost all the distribution lies within 3 standard deviations of the mean, that is, between 285 and 315, so your sketch should look like this:

Exercise 6b

Always draw sketches and check that your answer is sensible.

1 The masses of packages from a particular machine are normally distributed with mean 200 g and standard deviation 2 g. Find the probability that a randomly selected package from the production line of the machine weighs

   (i) less than 197 g,

   (ii) more than 200.5 g,

   (iii) between 198.5 g and 199.5 g.

2 The heights of boys at a particular age follow a normal distribution with mean 150.3 cm and standard deviation 5 cm. Find the probability that the height of a boy picked at random from this age group is

   (i) less than 153.2 cm,

   (ii) more than 158 cm,

   (iii) between 150 cm and 158 cm,

   (iv) more than 10 cm from the mean height.

3 $X \sim N(300, 25)$
   Find the probabilities represented by the shaded areas in the diagrams:

   (i)   (ii) 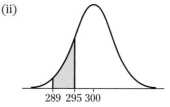   (iii)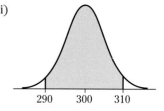

4 The random variable X is distributed normally with mean 50 and variance 20. Find

   (i) $P(X > 60.3)$,

   (ii) $P(X < 59.8)$.

5 $X \sim N(-8, 12)$. Find

   (i) $P(X < -9.8)$,

   (ii) $P(X > -8.2)$,

   (iii) $P(-7 \leqslant X \leqslant 0.5)$.

6 The masses of a certain type of cabbage are normally distributed with mean 1000 g and standard deviation 150 g.

   (a) Find the proportion of cabbages with a mass

       (i) greater than 850 g,

       (ii) between 750 g and 1290 g,

   (b) Estimate the number of cabbages in a batch of 800 with a mass less than 900 g or greater than 1375 g.

7 The number of hours of life of a certain type of torch battery is normally distributed with mean 150 and standard deviation 12. In a quality control test two batteries are chosen at random from a batch. If both batteries have a life less than 120 hours, the batch is rejected. Find the probability that the batch is rejected.

8   Cartons of milk from a particular supermarket are advertised as containing 1 litre of milk, but in fact the volume of the milk in a carton is normally distributed with mean 1012 ml and standard deviation 5 ml.

   (i) Find the probability that a randomly chosen carton contains more than 1010 ml.

   (ii) Find the probability that exactly 3 cartons in a sample of 10 cartons contain more than 1012 ml.

   (iii) Estimate how many cartons in a batch of 1000 cartons contain less than the advertised volume of milk.

9   The lifetime, in hours, of a certain make of electric light bulb is known to be normally distributed with mean 2000 and standard deviation 120.
   Estimate the probability that the lifetime of a bulb of this make will be

   (i) greater than 2150 hours,

   (ii) greater than 1910 hours,

   (iii) between 1850 hours and 2090 hours.

10   The weights of vegetable marrows supplied to retailers by a wholesaler have a normal distribution with mean 1.5 kg and standard deviation 0.02 kg. The wholesaler supplies three sizes of marrow:

   Size 1   under 1.48 kg
   Size 2   from 1.48 kg to 1.53 kg
   Size 3   over 1.53 kg

   Find, to three decimal places, the proportions of marrows in the three sizes.

11   A normal distribution has mean μ and standard deviation σ. 1000 observations are taken from the distribution. How many would you expect to be between $\mu - 1.5\sigma$ and $\mu + 1.5\sigma$?

12   An intelligence test used in a particular country has scores which are normally distributed with mean 100 and standard deviation 15. In a randomly selected group of 500 people sitting the test, estimate how many have a score

   (i) higher than 140,

   (ii) below 120,

   (iii) between 100 and 110,

   (iv) between 85 and 90.

FINDING z WHEN $\Phi(z)$ IS KNOWN

Notation

To find z when you know $\Phi(z)$ you have to read the tables 'in reverse'. The following notation is useful:

If $\Phi(z) = p$

$\Phi^{-1}(p)$ is read as 'phi to the minus 1 of p'

then $z = \Phi^{-1}(p)$

Look for p in the main body of the table, then 'read back' to get z. The method is illustrated below using an extract from the normal distribution tables.

z	0	1	2	3	4	5	6	7	8	9	1	2	3	4	5	6	7	8	9
															ADD				
1.5	0.9332	0.9345	0.9357	0.9370	0.9382	0.9394	0.9406	0.9418	0.9429	0.9441	1	2	4	5	6	7	8	10	11
1.6	0.9452	0.9463	0.9474	0.9484	0.9495	0.9505	0.9515	0.9525	0.9535	0.9545	1	2	3	4	5	6	7	8	9
1.7	0.9554	0.9564	0.9573	0.9582	0.9591	0.9599	0.9608	0.9616	0.9625	0.9633	1	2	3	4	4	5	6	7	8
1.8	0.9641	0.9649	0.9656	0.9664	0.9671	0.9678	0.9686	0.9693	0.9699	0.9706	1	1	2	3	4	4	5	6	6
1.9	0.9713	0.9719	0.9726	0.9732	0.9738	0.9744	0.9750	0.9756	0.9761	0.9767	1	1	2	2	3	4	4	5	5
2.0	0.9772	0.9778	0.9783	0.9788	0.9793	0.9798	0.9803	0.9808	0.9812	0.9817	0	1	1	2	2	3	3	4	4
2.1	0.9821	0.9826	0.9830	0.9834	0.9838	0.9842	0.9846	0.9850	0.9854	0.9857	0	1	1	2	2	2	3	3	4
2.2	0.9861	0.9864	0.9868	0.9871	0.9875	0.9878	0.9881	0.9884	0.9887	0.9890	0	1	1	1	2	2	2	3	3

(i) $\Phi(z) = 0.9406$

$\quad z = \Phi^{-1}(0.9406)$

$\quad\quad = 1.56$

Look for 0.9406 in the main body of the table. It occurs when $z = \mathbf{1.56}$.

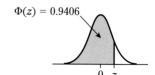

$\Phi(z) = 0.9406$

(ii) $\Phi(z) = 0.9579$

$\quad z = \Phi^{-1}(0.9579)$

$\quad\quad = 1.727$

Look for 0.9579 in the main body of the table. It does not appear, so look for the closest number *below* it. This is 0.9573 and it occurs when $z = \mathbf{1.72}$.

To get the digits 9579 you need to ADD 6 to 9573. Look at the far right section and find 6. It is in column 7. This means that $z = \mathbf{1.727}$.

(iii) $\Phi(z) = 0.9832$

$\quad z = \Phi^{-1}(0.9832)$

$\quad\quad = 2.125$

0.9832 is not in the main body of the table, but the closest number *below* it is 0.9830.

To get the digits 9832 you need to ADD 2 to 9830.

Look at the far right section; 2 occurs in columns 4, 5 and 6, giving $z = 2.124$ or $z = 2.125$ or $z = 2.126$. So each of these z values satisfies $\Phi(z) = 0.9832$.

Choose the middle value, say, so $z = \mathbf{2.125}$.

Example 6.9

Given $Z = N(0, 1)$, find a where

(i) $P(Z < a) = 0.9693$ (ii) $P(Z > a) = 0.3802$

(iii) $P(Z > a) = 0.7367$ (iv) $P(Z < a) = 0.0793$

Always draw a sketch.

Think carefully about where to position z.

Consider whether z is positive (in the right-hand or upper tail) or negative (in the left-hand or lower tail).

(i) $P(Z < a) = 0.9693$

$\quad\quad \Phi(a) = 0.9693$

$\quad\quad\quad a = \Phi^{-1}(0.9693)$

$\quad\quad\quad\quad = 1.87$

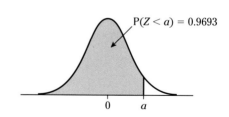

$P(Z < a) = 0.9693$

(ii) $P(Z > a) = 0.3802$

$$\Phi(a) = 1 - 0.3802 = 0.6198$$
$$a = \Phi^{-1}(0.6198)$$
$$= 0.305$$

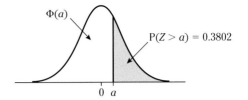

(iii) $P(Z > a) = 0.7367$
Since the probability is greater than 0.5,
a must be negative and therefore $-a$ is positive.
Using symmetry
$$\Phi(-a) = 0.7367$$
$$-a = \Phi^{-1}(0.7367)$$
$$= 0.633$$
$$a = -0.633$$

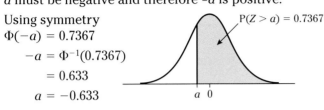

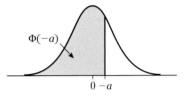

(iv) $P(Z < a) = 0.0793$
Since the probability is less than 0.5, a must be negative.
Using symmetry
$$\Phi(-a) = 1 - 0.0793 = 0.9207$$
$$-a = \Phi^{-1}(0.9207)$$
$$= 1.41$$
$$a = -1.41$$

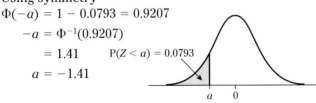

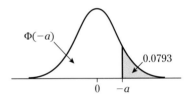

Critical values

Some values of z, corresponding to particular values of p, are important in later work and these are summarised in a table of **critical values**. This table is given in the examination and it is useful to become familiar with it as it could save you time.

Critical values for the normal distribution

If Z has a normal distribution with mean 0 and variance 1 then, for each value of p, the table gives the value of z such that $P(Z < z) = p$.

p	0.75	0.90	0.95	0.975	0.99	0.995	0.9975	0.999	0.9995
z	0.674	1.282	1.645	1.960	2.326	2.576	2.807	3.090	3.291

For example
- if $P(Z < z) = 0.95$, look up $p = 0.95$ to get $z = 1.645$
- if $P(Z > z) = 0.01$, then $P(Z < z) = 0.99$, so look up $p = 0.99$ to get $z = 2.326$

You have found $\Phi^{-1}(0.95)$.

Example 6.10

Given $Z = N(0, 1)$, find
(i) the upper quartile of the distribution
(ii) the interquartile range.

(i) If q is the upper quartile, then 75% of the distribution lies to the left of q, so

$$P(Z < q) = 0.75$$

In critical values table look up $p = 0.75$.

So $\Phi(q) = 0.75$

$q = 0.674$

The upper quartile is 0.674.

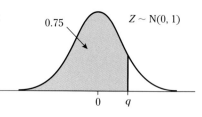

(ii) 25% of the distribution lies to the left of the lower quartile so, by symmetry, the lower quartile is -0.674.

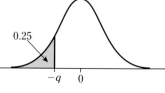

Interquartile range = upper quartile − lower quartile

$$= 0.674 - (-0.674)$$

$$= 1.348$$

Note that, by definition, the central 50% of the distribution lies between the lower and upper quartiles.

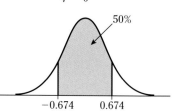

Exercise 6c

1 $Z \sim N(0, 1)$.

Find the value of z in each of the following:

(i)

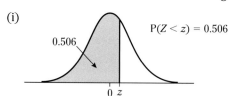

$P(Z < z) = 0.506$

(ii)

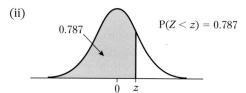

$P(Z < z) = 0.787$

(iii)

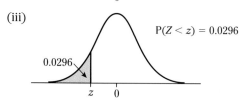

$P(Z < z) = 0.0296$

(iv)

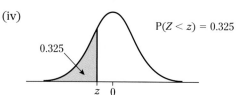

$P(Z < z) = 0.325$

(v)

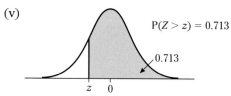

$P(Z > z) = 0.713$

(vi)

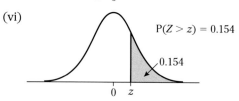

$P(Z > z) = 0.154$

2 $Z \sim N(0, 1)$.

Find the value of a in each of the following:

(i) $P(Z < a) = 0.9738$

(ii) $P(Z \leqslant a) = 0.2435$

(iii) $P(Z > a) = 0.82$

(iv) $P(Z > a) = 0.2351$

(v) $P(-a < Z < a) = 0.7$

USING THE TABLES IN REVERSE FOR *ANY* NORMAL VARIABLE X

If $X \sim \mathrm{N}(\mu, \sigma^2)$, to find an x value when you know a probability, first find the corresponding z value by using the normal distribution table in reverse. Then substitute this into $z = \dfrac{x - \mu}{\sigma}$ to find x.

Example 6.11

The time, X minutes, taken by pupils in a particular class to do their mathematics homework follows a normal distribution with mean 25 and standard deviation 7.

85% of the pupils take less than x minutes. Find the value of x.

X is the time taken, in minutes

$X \sim \mathrm{N}(25, 7^2)$ with $\mu = 25$ and $\sigma = 7$

Show the information on a sketch.

You are given

$$P(X < x) = 0.85$$

i.e. $P(Z < z) = 0.85$ where $z = \dfrac{x - 25}{7}$

$$z = \Phi^{-1}(0.85) = 1.036$$

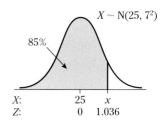

So $\dfrac{x - 25}{7} = 1.036$

$$x - 25 = 1.036 \times 7$$

$$x = 25 + 1.036 \times 7$$

$$= 32.252$$

So 85% of the pupils take less than 32.3 minutes (3 s.f.).

Example 6.12

Ada sells apples from her orchard at a roadside stall. The mass, in grams, of an apple from her orchard is a random variable with distribution $\mathrm{N}(104.8, 70)$. She describes the heaviest 5% of the apples as 'extra large'. Find the least mass of one of Ada's 'extra large' apples.

X is the mass, in grams, of an apple from the orchard.

$X \sim \mathrm{N}(104.8, 70)$ with $\mu = 104.8$, $\sigma = \sqrt{70}$

Take care here. You are told the variance, so $\sigma^2 = 70$ and $\sigma = \sqrt{70}$.

If m is the least mass of an 'extra large' apple

then $P(X > m) = 0.05$

i.e. $P(X < m) = 0.95$

So $P(Z < z) = 0.95$, where $z = \dfrac{m - 104.8}{\sqrt{70}}$

Now $z = \Phi^{-1}(0.95) = 1.645$

Input $\sqrt{70}$ directly on your calculator.

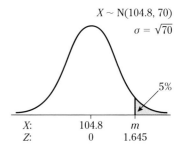

So $\dfrac{m - 104.8}{\sqrt{70}} = 1.645$

$$m = 104.8 + 1.645 \times \sqrt{70}$$

$$= 118.56$$

The least mass of an extra large apple is 119 g (3 s.f.).

Example 6.13

Tyre pressures on a certain type of car independently follow a normal distribution with mean 1.9 bars and standard deviation 0.15 bars.

(i) Find the probability that all four tyres on a car of this type have pressures between 1.82 bars and 1.92 bars.

(ii) Safety regulations state that the pressures must be between $1.9 - b$ bars and $1.9 + b$ bars. It is known that 80% of tyres are within these safety limits. Find the safety limits.

Cambridge Paper 6 Q6 J05

X is pressure in bars.
$X \sim N(1.9, 0.15^2)$ with $\mu = 1.9$, $\sigma = 0.15$

(i) $P(1.82 < X < 1.92)$

$$= P\left(\frac{1.82 - 1.9}{0.15} < Z < \frac{1.92 - 1.9}{0.15}\right)$$

$$= P(-0.533 < Z < 0.133)$$

$$= \Phi(0.133) - (1 - \Phi(0.533))$$

$$= 0.5529 - (1 - 0.7029)$$

$$= 0.2558$$

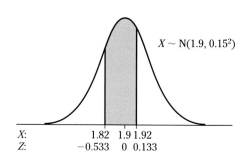

So, P(all four tyres have pressures between 1.82 bars and 1.92 bars) $= (0.2558)^4 = 0.004281... = 0.00428$ (3 s.f.)

Showing the information on a sketch will help you to see the probabilities

(ii) The central 80% of the distribution lies between $1.9 - b$ and $1.9 + b$.

From the diagram, you can see that 90% of the area is to the left of $(1.9 + b)$.

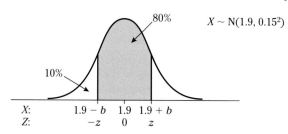

So $P(X < 1.9 + b) = 0.9$

i.e. $P(Z < z) = 0.9$

$\Phi(z) = 0.9$

$z = \Phi^{-1}(0.9) = 1.282$

Now $z = \dfrac{x - \mu}{\sigma} = \dfrac{(1.9 + b) - 1.9}{0.15} = \dfrac{b}{0.15}$

So $\dfrac{b}{0.15} = 1.282$

$b = 0.1923$

$1.9 - 0.1923 = 1.71$ (3 s.f.)

$1.9 + 0.1923 = 2.09$ (3 s.f.)

So the safety limits are 1.71 bars and 2.09 bars.

Example 6.14

Bags of flour packed by a particular machine have masses that are normally distributed with mean 508 g and standard deviation 4 g. It is found that 2% of the bags are rejected for being underweight.

  (i) Find the minimum mass for a bag to be accepted.

 (ii) Six bags of flour are selected at random from the production line. Find the probability that at most one of these bags is underweight.

  (i) X is the mass, in g, of a bag of flour.

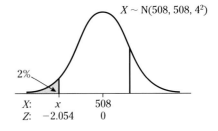

$$X \sim N(508, 4^2) \qquad \mu = 508, \sigma = 4$$

$$P(X < x) = 0.02$$

i.e. $P(Z < z) = 0.02$ where $z = \dfrac{x - 508}{4}$

$$\Phi(-z) = 1 - 0.02 = 0.98$$

$$-z = \Phi^{-1}(0.98) = 2.054$$

$$z = -2.054$$

So, $\dfrac{x - 508}{4} = -2.054$

$$x = 508 + (-2.054) \times 4$$

$$= 499.784$$

$$= 499.8 \ (1 \ \text{d.p.})$$

The minimum mass for a bag to be accepted is 499.8 g.

 (ii) P(bag is underweight) $= 0.02$.

Let W be the number of bags in 6 that are underweight.

> This is a binomial situation where $P(X = x) = \binom{n}{x}p^x q^{n-x}$ (page **160**).

$W \sim B(6, 0.02)$ with $n = 6, p = 0.02, q = 0.98$.

P(at most one bag is underweight)

$$= P(W = 0) + P(W = 1)$$

$$= 0.98^6 + \binom{6}{1} \times 0.02^1 \times 0.98^5$$

$$= 0.9943\ldots$$

$$= 0.994 \ (3 \ \text{s.f.})$$

Exercise 6d

1 Find x in each of the following.

(i) $X \sim N(60, 5^2)$

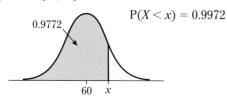

P($X < x$) = 0.9972

(ii) $X \sim N\left(5, \frac{4}{9}\right)$

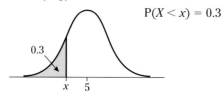

P($X < x$) = 0.3

(iii) $X \sim N(200, 6^2)$

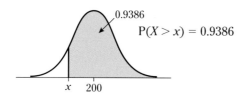

P($X > x$) = 0.9386

(iv) $X \sim N(0, 4)$

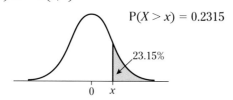

P($X > x$) = 0.2315

2 The heights of female students at a particular college are normally distributed with mean 169 cm and standard deviation 9 cm.

(i) Given that 80% of these female students have a height less than h cm, find the value of h.

(ii) Given that 60% of these female students have a height greater than s cm, find the value of s.

3 The masses of lettuces sold at a market stall are normally distributed with mean 600 g and standard deviation 20 g.

(i) Find the mass exceeded by 10% of the lettuces.

(ii) 5% of the lettuces have a mass less than M g. Find M.

4 $X \sim N(400, 64)$.

(i) Find the limits within which the central 95% of the distribution lies.

(ii) Find the interquartile range of the distribution.

5 The lengths of metal rods produced by a machine are normally distributed with mean 120 cm and standard deviation 10 cm.

(i) Find the probability that a rod selected at random has a mass within 5 cm of the mean.

Rods shorter than L cm are rejected.

(ii) Estimate the value of L if 1% of all rods are rejected.

6 The lifetime, in hours, of a certain make of batteries follows a normal distribution with mean 160 and standard deviation 30. Calculate the interval, symmetrical about the mean, within which 75% of the battery lifetimes lie.

7 The times for a certain car journey have a normal distribution with mean 100 minutes and standard deviation 7 minutes. Journey times are classified as follows:

'short'	(the shortest 33% of times)
'long'	(the longest 33% of times)
'standard'	(the remaining 34% of times)

(i) Find the probability that a randomly chosen car journey takes between 85 and 100 minutes.

(ii) Find the least and greatest times for the 'standard' journey.

Cambridge Paper 61 Q3 N09

FINDING μ OR σ OR BOTH

Example 6.15

The length of Paulo's lunch break follows a normal distribution with mean μ minutes and standard deviation 5 minutes. On one day in four, on average, his lunch break lasts for more than 52 minutes.

(i) Find the value of μ.

(ii) Find the probability that Paulo's lunch break lasts for between 40 and 46 minutes on every one of the next four days.

Cambridge Paper 6 Q5 N04

X is the time, in minutes, that the lunch break lasts.

$X \sim N(\mu, 5^2)$ μ is unknown, $\sigma = 5$

(i) $P(X > x) = \frac{1}{4} = 0.25$

   i.e. $P(Z > z) = 0.25$, where $z = \dfrac{52 - \mu}{5}$

$$\Phi(z) = 1 - 0.25 = 0.75$$

$$z = \Phi^{-1}(0.75) = 0.674$$

So $\dfrac{52 - \mu}{5} = 0.674$

$$52 - \mu = 0.674 \times 5$$

$$\mu = 52 - 0.674 \times 5$$

$$= 48.63$$

$$= 48.6 \text{ (3 s.f.)}$$

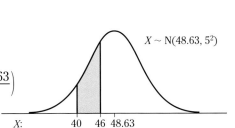

$X \sim N(\mu, 5^2)$

0.25

X: μ 52
Z: 0 0.674

(ii) You now know that

$X \sim N(48.63, 5^2)$ with $\mu = 48.63$, $\sigma = 5$

$P(40 < X < 46) = P\left(\dfrac{40 - 48.63}{5} < Z < \dfrac{46 - 48.63}{5}\right)$

$$= P(-1.726 < Z < -0.526)$$

$$= \Phi(1.726) - \Phi(0.526)$$

$$= 0.9578 - 0.7005$$

$$= 0.2573$$

$$= 0.257 \text{ (3 s.f.)}$$

$X \sim N(48.63, 5^2)$

X: 40 46 48.63
Z: -1.726 -0.526 0

P(between 40 and 46 minutes for next four days)

$$= (0.2573)^4$$

$$= 0.004382\ldots$$

$$= 0.00438 \text{ (3 s.f.)}$$

Use 0.2573 here, not the rounded value of 0.257. If you use 0.257 you will get 0.00436 (3 s.f.). This does not give the required answer and you are likely to lose accuracy marks because of premature rounding.

Remember that you must work to at least 4 figures if your final answer is to be given to 3 significant figures.

Example 6.16

The random variable X is normally distributed with mean 100 and $P(X < 106) = 0.8849$. Find the standard deviation.

$X \sim \mathrm{N}(100, \sigma^2)$ $\qquad$ $\mu = 100$, σ is unknown

$\qquad$ $\mathrm{P}(X < 106) = 0.8849$

i.e. $\qquad$ $\mathrm{P}(Z < z) = 0.8849$

so $\qquad$ $\Phi(z) = 0.8849$

$\qquad\qquad z = \Phi^{-1}(0.8849)$

$\qquad\qquad\quad = 1.2$

Now $z = \dfrac{x - \mu}{\sigma} = \dfrac{106 - 100}{\sigma} = \dfrac{6}{\sigma}$

so $\qquad 1.2 = \dfrac{6}{\sigma}$

$\qquad 1.2\sigma = 6$

$\qquad\quad \sigma = \dfrac{6}{1.2} = 5$

The standard deviation is 5.

Example 6.17

The random variable X is normally distributed. The mean is twice the standard deviation. It is given that $\mathrm{P}(X > 5.2) = 0.9$. Find the standard deviation.

Cambridge Paper 6 Q3(a) J07

$X \sim \mathrm{N}(\mu, \sigma^2)$ $\qquad$ $\mu = 2\sigma$

$\mathrm{P}(X > 5.2) = 0.9$

$\qquad \Phi(-z) = 0.9$

$\qquad\quad -z = \Phi^{-1}(0.9)$

$\qquad\quad -z = 1.282$

$\qquad\quad\ \ z = -1.282$

Now $z = \dfrac{x - \mu}{\sigma} = \dfrac{5.2 - 2\sigma}{\sigma}$

so $\qquad \dfrac{5.2 - 2\sigma}{\sigma} = -1.282$

$\qquad 5.2 - 2\sigma = -1.282\sigma$

$\qquad\qquad\ 5.2 = 0.718\sigma$

$\qquad\qquad\ \ \sigma = \dfrac{5.2}{0.718}$

$\qquad\qquad\qquad = 7.243\ldots$

$\qquad\qquad\qquad = 7.24 \text{ (3 s.f.)}$

You need to consider where to place 5.2 on the diagram.

Notice that 90% of the area is to the right of 5.2, so 5.2 must be below the mean and z must be negative.

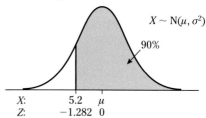

Sometimes you have to find the values of both μ and σ.

Example 6.18

The mass in kg of boxes of oranges is normally distributed such that 30% of them have a mass greater than 4.00 kg and 20% have a mass greater than 4.53 kg. Find the mean and standard deviation of the masses.

X is the mass in kg of a box of oranges.

$X \sim \text{N}(\mu, \sigma^2)$ with both μ and σ unknown

$\text{P}(X > 4.00) = 0.3$

$\quad \text{P}(Z > z_1) = 0.3$ where $z_1 = \dfrac{4.00 - \mu}{\sigma}$

$\quad\quad \Phi(z_1) = 0.7$

$\quad\quad\quad z_1 = \Phi^{-1}(0.7)$

$\quad\quad\quad\quad = 0.524$

So $\quad \dfrac{4.00 - \mu}{\sigma} = 0.524$

$\quad\quad 4.00 - \mu = 0.524\sigma$

$\quad\quad\quad 4.00 = \mu + 0.524\sigma \quad\quad (1)$

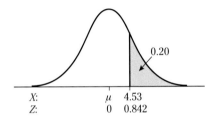

$\text{P}(X > 4.53) = 0.2$

$\text{P}(Z > z_2) = 0.2$ where $z_2 = \dfrac{4.53 - \mu}{\sigma}$

$\quad \Phi(z_2) = 0.8$

$\quad\quad z_2 = \Phi^{-1}(0.8)$

$\quad\quad\quad = 0.842$

So $\quad \dfrac{4.53 - \mu}{\sigma} = 0.842$

$\quad\quad 4.53 - \mu = 0.842\sigma$

$\quad\quad\quad 4.53 = \mu + 0.842\sigma \quad\quad (2)$

Now solve the simultaneous equations to find μ and σ.

$\quad\quad\quad\quad 4.53 = \mu + 0.842\sigma \quad\quad (2)$

$\quad\quad\quad\quad 4.00 = \mu + 0.524\sigma \quad\quad (1)$

$(2) - (1) \quad 0.53 = 0.318\sigma$

Store 1.666... in your calculator memory.

$\quad\quad\quad \sigma = \dfrac{0.53}{0.318} = 1.666... = 1.67 \ (3 \text{ s.f.})$

Substitute for σ in (1)

$\quad\quad 4.00 = \mu + 0.524 \times 1.666...$

$\quad\quad\quad \mu = 4.00 - 0.524 \times 1.666...$

$\quad\quad\quad\quad = 3.126... = 3.13 \ (3 \text{ s.f.})$

If 1.666... is not still on your calculator display, recall it from the memory. Do not use 1.67 as this will result in loss of accuracy.

So $\quad \mu = 3.13$ and $\sigma = 1.67$

Example 6.19

The random variable X is normally distributed with mean μ and variance σ^2. It is given that $\text{P}(X > 200) = 0.0166$ and $\text{P}(X > 170) = 0.8461$. Find μ and σ.

$X \sim \text{N}(\mu, \sigma^2)$

$\text{P}(X > 200) = 0.0166$

$\quad \text{P}(Z > z_1) = 0.0166$ where $z_1 = \dfrac{200 - \mu}{\sigma}$

$\quad\quad \Phi(z_1) = 1 - 0.0166 = 0.9834$

$\quad\quad\quad z_1 = \Phi^{-1}(0.9834) = 2.13$

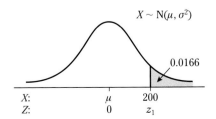

So $\dfrac{200 - \mu}{\sigma} = 2.13$

$200 - \mu = 2.13\sigma$ (1)

$P(X > 170) = 0.8461$

$P(Z > z_2) = 0.8461$ where $z_2 = \dfrac{170 - \mu}{\sigma}$

$\Phi(-z_2) = 0.8461$

$-z_2 = \Phi^{-1}(0.8461) = 1.02$

$z_2 = -1.02$

So $\dfrac{170 - \mu}{\sigma} = -1.02$

$170 - \mu = -1.02\sigma$ (2)

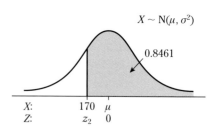

$X \sim N(\mu, \sigma^2)$

0.8461

Now solve the simultaneous equations to find μ and σ.

$200 - \mu = 2.13\sigma$ (1)

$170 - \mu = -1.02\sigma$ (2)

(1) − (2) $30 = 3.15\sigma$

Take care here:
$2.13\sigma - (-1.02\sigma) = 3.15\sigma$

$\sigma = \dfrac{30}{3.15} = 9.523...$

Store 9.523... in calculator memory.

Sub in (1) $200 - \mu = 2.13 \times 9.523...$

$200 - \mu = 20.28...$

$\mu = 200 - 20.28... = 179.71...$

So $\mu = 180$ (3 s.f.) and $\sigma = 9.52$ (3 s.f.).

Exercise 6e

You are advised to draw sketches and check that your answer is reasonable.

1 The random variable X is normally distributed with standard deviation 25.
If $P(X < 27.5) = 0.3085$, find the mean.

2 The random variable X is normally distributed with mean 45 and standard deviation σ. Given that $P(X > 51) = 0.288$, find the value of σ.

3 The volume of orange juice in cartons sold by a particular company is normally distributed with mean 333 ml. It is known that 20% of the cans contain more than 340 ml.

(i) Find the standard deviation of the volume of orange juice in a can.

(ii) Find the percentage of cans that contain less than 330 ml.

4 The random variable X is distributed as $N(\mu, 12)$. It is given that $P(X > 32) = 0.8438$.

(i) Find the value of μ.

(ii) Find the probability that a random observation of X lies between 34.5 and 35.3.

5 The heights, in metres, of 500 people are normally distributed with standard deviation 0.080. Given that the heights of 129 of these people are greater than the mean height, but less than 1.806 m, estimate the mean height.

Hint: Draw a sketch.

6 The masses of boxes of apples are normally distributed such that 20% of the boxes are heavier than 5.08 kg and 15% of the boxes are heavier than 5.62 kg.

Estimate the mean and standard deviation of the masses.

7 The random variable X has a normal distribution with mean μ and standard deviation σ. Given that $P(X > 80) = 0.0113$ and $P(X < 30) = 0.0287$, find μ and σ.

8 Metal rods produced by a machine have lengths that are normally distributed. It is known that 2% of the rods are rejected as being too short and 5% are rejected as being too long.

   (i) Given that the least and greatest acceptable lengths of the rods are 6.32 cm and 7.52 cm, calculate the mean and variance of the lengths of the rods.

   (ii) Ten rods are chosen at random from a batch produced by the machine. Find the probability that exactly three of them are rejected for being too long.

9 The random variable X is distributed as $N(\mu, \sigma^2)$.
$P(X < 35) = 0.2$ and $P(35 < X < 45) = 0.65$.
Find μ and σ.

10 (i) The height of sunflowers follows a normal distribution with mean 112 cm and standard deviation 17.2 cm. Find the probability that the height of a randomly chosen sunflower is greater than 120 cm.

   (ii) When a new fertiliser is used, the height of sunflowers follows a normal distribution with mean 115 cm. Given that 80% of the heights are now greater than 103 cm, find the standard deviation.

                      Cambridge Paper 6 Q3 J03

11 A farmer cuts hazel twigs to make into bean poles to sell at the market. He says that the sticks are each 240 cm long but in fact the lengths of the sticks are normally distributed such that 55% of the sticks are longer than 240 cm and 10% are longer than 250 cm.

Find the probability that a randomly selected stick is shorter than 235 cm.

12 The diameters of bolts produced by a particular machine follow a normal distribution with mean 1.34 cm and standard deviation 0.04 cm. A bolt is rejected if its diameter is less than 1.24 cm or more than 1.40 cm.

   (i) Find the percentage of bolts which are accepted.

The setting of the machine is altered so that the mean diameter changes but the standard deviation remains the same. With the new setting, 3% of the bolts are rejected because they are too large in diameter.

   (ii) Find the new mean diameter of bolts produced by the machine.

   (iii) Find the percentage of bolts that are now rejected because they are too small in diameter.

13 Tea is sold in packages marked 750 g. The masses of the packages are normally distributed with mean 760 g. The probability that a package weighs more than 750 g is 0.975. Find the standard deviation of the distribution.

14 In a normal distribution, 4% of the distribution is less than 53 and 97% of the distribution is less than 65.

   (i) Find the mean and standard deviation of the distribution.

   (ii) Find the interquartile range of the distribution.

15 A certain make of car tyre can be safely used for 25 000 km on average before it is replaced. The manufacturer guarantees to pay compensation to anyone whose tyre does not last for 20 000 km. It is thought that 5% of all tyres sold will qualify for compensation. The distance travelled, X km, before a tyre is replaced has a normal distribution. Find the standard deviation of X.

16 A machine dispenses raisins into bags so that the mass of raisins in a bag is normally distributed.

   (i) The standard deviation is 1.25 g. It is found that 2.5% of bags contain less than 826 g of raisins. Find the mean mass of raisins in a bag.

   (ii) After the machine is serviced, the mean mass of raisins in a bag is 828.1 g and 0.1% of bags contain more than 830 g. Find the standard deviation of the mass of raisins in a bag after the service.

17 A machine is used to fill cans of soup with a nominal volume of 0.5 litres. The quantity of soup actually delivered is normally distributed with mean μ litres and standard deviation σ litres. It is required that no more than 1% of cans should contain less than the nominal volume.

   (i) Find the least value of μ which will comply with the requirement when $\sigma = 0.003$.

   (ii) Find the greatest value of σ which will comply with the requirement when $\mu = 0.506$.

18 In a large consignment of packets of sugar, it is found that 5% have a mass less than 515 g and 2% have a mass less than 510 g. The masses of packets of sugar are normally distributed. Estimate the mean and the standard deviation of the mass of sugar in a packet.

19 The speeds of cars passing a checkpoint on a certain motorway follow a normal distribution. Observations show that 95% of the cars passing the checkpoint are travelling at less than 136 km/h and 10% are travelling at less than 88 km/h.

   (i) Find the average speed of cars passing the checkpoint.

   (ii) Find the proportion of cars travelling at more than 112 km/h.

20 The random variable X has a normal distribution with mean μ and standard deviation σ.

   (i) If $2\mu = 1.5\sigma$, find $P(X < 3\mu)$.

   (ii) If $P(X > \frac{1}{3}\mu) = 0.7257$, express μ in terms of σ.

THE NORMAL APPROXIMATION TO THE BINOMIAL DISTRIBUTION

Under certain circumstances the normal distribution can be used as an approximation to the binomial distribution. One practical advantage is that the calculations for finding probabilities are less tedious to perform.

The diagrams below illustrate the distribution $B(n, p)$ for $p = 0.2$ and $p = 0.5$, for various values of n. In each case a vertical line graph has been drawn and, to make the comparison easier to see, a curve has been superimposed on each.

$p = 0.5$

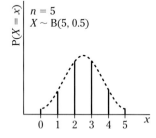
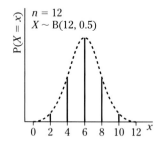
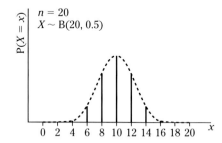

When $p = 0.5$, the binomial distribution is symmetrical. Even when n is quite small, the distribution takes on the characteristic bell shape of a normal distribution.

$p = 0.2$

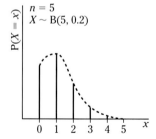
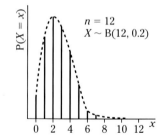
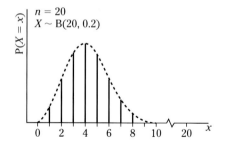

When $p = 0.2$, for small values of n the distribution has a positive skew, but when $n = 20$ the distribution is almost symmetrical and bell-shaped.

You will recall from Chapter 5 (page **168**) that when the discrete random variable X follows a **binomial distribution**, where $X \sim \text{B}(n, p)$, then

$$\mu = \text{E}(X) = np \text{ and } \sigma^2 = \text{Var}(X) = npq \text{ where } q = 1 - p.$$

Under certain conditions a normal distribution with the mean np and variance npq can be used as an approximation for the binomial distribution.

The following rule can be used.

If $\qquad X \sim \text{B}(n, p)$

and n is large enough to ensure that $np > 5$ and $nq > 5$, where $q = 1 - p$,

then $\qquad X \sim \text{N}(np, npq)$ with $\mu = np$ and $\sigma = \sqrt{npq}$.

Note that the farther p is from 0.5, the larger n needs to be for the conditions to be satisfied.

Continuity corrections

The following example compares probabilities obtained using a binomial distribution and a normal distribution. It also illustrates the use of a **continuity correction** which is needed when using a continuous distribution (the normal distribution) as an approximation to a discrete distribution (the binomial distribution).

Example 6.20

Find the probability of obtaining 4, 5, 6 or 7 heads when a fair coin is tossed 12 times
  (i) using a binomial distribution,
 (ii) using a normal approximation to the binomial distribution.

X is the number of heads in 12 tosses.
$X \sim \text{B}(12, 0.5)$ with $n = 12, p = 0.5$

  (i) Using the binomial distribution:

$$\text{P}(X = 4) = \binom{12}{4}(0.5)^4(0.5)^8 = \binom{12}{4}(0.5)^{12} = 0.1208\ldots$$

$$\text{P}(X = 5) = \binom{12}{5}(0.5)^5(0.5)^7 = \binom{12}{5}(0.5)^{12} = 0.1933\ldots$$

$$\text{P}(X = 6) = \binom{12}{6}(0.5)^6(0.5)^6 = \binom{12}{6}(0.5)^{12} = 0.2255\ldots$$

$$\text{P}(X = 7) = \binom{12}{7}(0.5)^7(0.5)^5 = \binom{12}{7}(0.5)^{12} = 0.1933\ldots$$

Binomial probabilities, page **160**
$\text{P}(X = x) = \binom{n}{x}p^x q^{n-x}$

So $\text{P}(4 \leqslant X \leqslant 7)$

$\qquad = 0.1208\ldots + 0.1933\ldots + 0.2255\ldots + 0.1933\ldots$

$\qquad = 0.733 \text{ (3 d.p.)}$

(ii) The diagram below shows the probability distribution for $X \sim B(12, 0.5)$. The vertical lines have been replaced by rectangles to help to illustrate the intention to use a continuous distribution as an approximation to a discrete one. The required binomial probability is represented by the sum of the areas of the shaded rectangles.

First check the conditions for a normal approximation:

$np = 12 \times 0.5 = 6$, so $np > 5$

$nq = 12 \times 0.5 = 6$, so $nq > 5$

As both the conditions are satisfied, use the normal approximation

$X \sim N(np, npq)$ with $np = 6$ and $npq = 12 \times 0.5 \times 0.5 = 3$

So $X \sim N(6, 3)$ with $\mu = 6$ and $\sigma = \sqrt{3}$

Superimposing the curve which is approximately $N(6, 3)$, the probability of obtaining 4, 5, 6 or 7 heads is found by considering the area under this normal curve from $x = 3.5$ to $x = 7.5$.

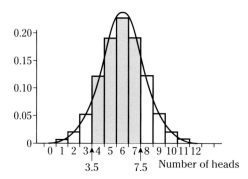

$P(4 \leqslant X \leqslant 7)$ transforms to $P(3.5 < X < 7.5)$ using a **continuity correction**.

Writing the symbol $\rightarrow$ to represent 'transforms to'

$P(4 \leqslant X \leqslant 7) \rightarrow P(3.5 < X < 7.5)$

> Input $\sqrt{3}$ using $\sqrt{\ }$ on your calculator. If you use a rounded value you will lose accuracy.

$$= P\left(\frac{3.5 - 6}{\sqrt{3}} < Z < \frac{7.5 - 6}{\sqrt{3}}\right)$$

$$= P(-1.443 < Z < 0.866)$$

$$= \Phi(0.866) - (1 - \Phi(1.443))$$

$$= 0.8067 - (1 - 0.9255)$$

$$= 0.732 \text{ (3 d.p.)}$$

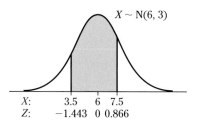

Note that the probabilities found by the two different methods compare well and the working for part (ii) is quicker to perform.

The approximation is good because, although n is not very large, $p = 0.5$.

More about continuity corrections

Continuity corrections sometimes cause difficulties, so these are considered in more detail using the diagram for the distribution of the number of heads when a coin is tossed 12 times.

If you want the probability that there are **at most** three heads
i.e. $P(X \leqslant 3)$, then consider $P(X < 3.5)$.

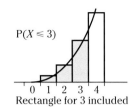

$P(X \leqslant 3)$

Rectangle for 3 included

If you want the probability that there are **fewer** than three heads,
i.e. $P(X < 3)$, then consider $P(X < 2.5)$.

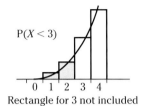

$P(X < 3)$

Rectangle for 3 not included

If you want the probability that there are exactly three heads,
i.e. $P(X = 3)$, then consider $P(2.5 < X < 3.5)$

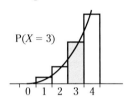

$P(X = 3)$

Here are some more examples.

$P(5 \leqslant X \leqslant 8) \rightarrow P(4.5 < X < 8.5)$ 5, 6, 7 or 8 heads
$P(5 < X \leqslant 8) \rightarrow P(5.5 < X < 8.5)$ 6, 7 or 8 heads
$P(5 \leqslant X < 8) \rightarrow P(4.5 < X < 7.5)$ 5, 6 or 7 heads
$P(5 < X < 8) \rightarrow P(5.5 < X < 7.5)$ 6 or 7 heads

Exercise 6f

Write down the continuity corrections needed
when the following binomial probabilities are
found using the normal approximation to the
binomial distribution.

1 $P(3 \leqslant X \leqslant 9)$

2 $P(3 < X < 9)$

3 $P(10 < X \leqslant 24)$

4 $P(2 \leqslant X < 8)$

5 $P(X > 24)$

6 $P(X \geqslant 76)$

7 $P(X < 109)$

8 $P(X \leqslant 45)$

9 $P(X = 67)$

10 $P(X = 20)$

Example 6.21

In a sack of mixed grass seeds, the probability that a seed is ryegrass is 0.35. Find the
probability that in a random sample of 400 seeds from the sack
  (i)   fewer than 120 are ryegrass seeds,
  (ii)   at least 125 but at most 132 are ryegrass seeds,
  (iii)   more than 160 are ryegrass seeds.

X is the number of ryegrass seeds in a sample of 400 seeds.

$X \sim B(400, 0.35)$ with $n = 400$, $p = 0.35$, $q = 0.65$

Define X as binomial first, then check that the conditions are suitable before defining the normal approximation.

To see whether a normal approximation is suitable, check the values of np and nq.

$np = 400 \times 0.35 = 140 > 5$

$nq = 400 \times 0.65 = 260 > 5$

Since both conditions are satisfied, use the normal approximation.

$X \sim N(np, npq)$
  with $np = 140$ and $npq = 400 \times 0.35 \times 0.65 = 91$

So $X \sim N(140, 91)$ $\qquad \mu = 140, \sigma = \sqrt{91}$

(i) $P(X < 120) \rightarrow P(X < 119.5)$ $\qquad$ *Apply a continuity correction*

$\qquad = P\left(Z < \dfrac{119.5 - 140}{\sqrt{91}}\right)$ $\qquad$ *Input $\sqrt{91}$ using $\sqrt{}$ on your calculator.*

$\qquad = P(Z < -2.149)$

$\qquad = 1 - \Phi(2.149)$

$\qquad = 0.0158$

(ii) $P(125 \leqslant X \leqslant 132) \rightarrow P(124.5 < X < 132.5)$

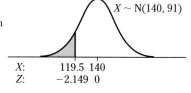

$\qquad = P\left(\dfrac{124.5 - 140}{\sqrt{91}} < Z < \dfrac{132.5 - 140}{\sqrt{91}}\right)$

$\qquad = P(-1.625 < Z < -0.786)$

$\qquad = \Phi(1.625) - \Phi(0.786)$ $\qquad$ *Result 2 page **180**.*

$\qquad = 0.9479 - 0.7841$

$\qquad = 0.1638$

$\qquad = 0.164$ (3 s.f.)

(iii) $P(X > 160) \rightarrow P(X > 160.5)$

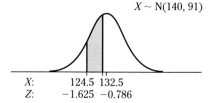

$\qquad = P\left(Z > \dfrac{160.5 - 140}{\sqrt{91}}\right)$

$\qquad = P(Z > 2.149)$

$\qquad = 1 - \Phi(2.149)$

$\qquad = 1 - 0.9842$

$\qquad = 0.0158$

Example 6.22

(i) A garden shop sells polyanthus plants in boxes, each box containing the same number of plants. The number of plants per box which produce yellow flowers has a binomial distribution with mean 11 and variance 4.95.

    (a) Find the number of plants per box.

    (b) Find the probability that a box contains exactly 12 plants which produce yellow flowers.

(ii) Another garden shop sells polyanthus plants in boxes of 100. The shop's advertisement states that the probability of any polyanthus plant producing a pink flower is 0.3. Use a suitable approximation to find the probability that a box contains fewer than 35 plants which produce pink flowers. $\qquad$ Cambridge Paper 6 Q7 J02

(i) X is 'the number of plants producing yellow flowers'.

$X \sim \text{B}(n, p)$ n and p are both unknown

(a) You are given that

$$np = 11 \qquad (1)$$
$$npq = 4.95 \qquad (2)$$

Recall: In a binomial distribution, mean $= np$ and variance $= npq$ where $q = 1 - p$. See page **168**.

Substitute np from (1) into (2)

$$11q = 4.95$$
$$q = \frac{4.95}{11} = 0.45$$

So $p = 1 - 0.45 = 0.55$

Substitute for p in (1)

$$n \times 0.55 = 11$$
$$n = \frac{11}{0.55} = 20$$

There are 20 plants in a box.

(b) $X \sim \text{B}(20, 0.55)$ with $n = 20$, $p = 0.55$, $q = 0.45$

$$P(X = 12) = \binom{20}{12} \times 0.55^{12} \times 0.45^{8}$$
$$= 0.1623\ldots = 0.162 \text{ (3 s.f.)}$$

Recall from page **160**.
$$P(X = x) = \binom{n}{x}p^x q^{n-x}$$

(ii) Y is 'the number of plants producing a pink flower in a box of 100'.

$Y \sim \text{B}(100, 0.3)$ with $n = 100$, $p = 0.3$, $q = 0.7$

Check the conditions for a normal approximation to the binomial distribution.

Now $np = 100 \times 0.3 = 30 > 5$
$nq = 100 \times 0.7 = 70 > 5$

Since $np > 5$ and $nq > 5$, use the normal approximation.

$Y \sim \text{N}(np, npq)$ where $np = 30$ and $npq = 100 \times 0.3 \times 0.7 = 21$

so $Y \sim \text{N}(30, 21)$ with $\mu = 30$ and $\sigma = \sqrt{21}$

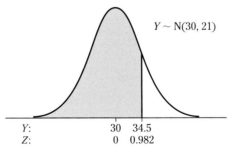

$Y \sim \text{N}(30, 21)$

Y: 30 34.5
Z: 0   0.982

$$P(Y < 35) \rightarrow P(Y < 34.5)$$

Apply a continuity correction

$$= P\left(Z < \frac{34.5 - 30}{\sqrt{21}}\right)$$
$$= P(Z < 0.9819\ldots)$$
$$= \Phi(0.982)$$
$$= 0.8370$$
$$= 0.837 \text{ (3 s.f.)}$$

Round to 3 decimal places and find $\Phi(0.982)$

Exercise 6g

1 A fair cubical die is thrown 120 times. Using a suitable approximation, find the probability of obtaining at least 24 sixes.

2 The random variable X follows a binomial distribution with $n = 25$ and $p = 0.38$.

   (i) Find the mean and variance of X.

   (ii) Verify that the distribution can be approximated by a normal distribution.

   (iii) Use the normal approximation to find

      (a) $P(10 \leqslant X \leqslant 15)$ (b) $P(X = 12)$

      (c) $P(X < 7)$ (d) $P(X \geqslant 9)$

3 On average 1 in 10 of the chocolates produced in a factory are mis-shapes. In a random sample of 1000 chocolates, find the probability that

   (i) fewer than 80 are mis-shapes,

   (ii) between 90 and 115 (inclusive) are mis-shapes,

   (iii) 120 or more are mis-shapes.

4 When Alex tries to send a fax, the probability that he can successfully send it is 0.85 and each attempt is independent of all other attempts. He tries to send 50 faxes. Use a suitable approximation to find the probability that he can successfully send at least 46 faxes.

5 At a particular hospital, records show that each day, on average, 80% of people keep their appointment at the outpatients' clinic.

   (i) Find the probability that in a random sample of 10 patients, more than 7 keep their appointment.

   (ii) Using a suitable approximation, find the probability that on a day when 200 appointments have been booked

      (a) more than 170 patients keep their appointments,

      (b) at least 155 patients keep their appointments.

6 The random variable X is distributed as B(200, 0.7). Use the normal approximation to the binomial distribution to find

   (i) $P(X \geqslant 130)$ (ii) $P(136 \leqslant X < 148)$

   (iii) $P(X < 142)$ (iv) $P(X = 152)$

7 One-fifth of a certain population has a minor eye defect. Use the normal distribution as an approximation to the binomial distribution to estimate the probability that the number of people with this eye defect is

   (i) more than 20 in a random sample of 100 people,

   (ii) exactly 20 in a random sample of 100 people,

   (iii) more than 200 in a random sample of 1000 people.

8 A certain variety of flower seed is sold in packets containing a large number of seeds. It is claimed on the packet that 40% will bloom white and 60% will bloom red and this may be assumed to be accurate.

   (i) Five seeds are planted. Find the probability that

      (a) exactly three will bloom white,

      (b) at least one will bloom white.

   (ii) One hundred seeds are planted. Use a suitable approximation to estimate the probability that at least 30 but at most 45 will bloom white.

9 A certain tribe is distinguished by the fact that 45% of the males have six toes on their right foot. Find the probability that, in a group of 200 males from the tribe, more than 97 have six toes on their right foot.

10 The random variable X follows a binomial distribution with $n = 12$ and $p = 0.42$.

   (i) Use the binomial distribution to calculate, to 4 significant figures, the probability that $X \leqslant 3$.

   (ii) Use the normal distribution as an approximation to the binomial distribution to calculate, to 4 significant figures, the probability that $X \leqslant 3$.

   (iii) Calculate the percentage error when using the normal approximation.

11   Kamal has 30 hens. The probability that any hen lays an egg on any day is 0.7. Hens do not lay more than one egg per day, and the days on which a hen lays an egg are independent.

(i) Calculate the probability that, on any particular day, Kamal's hens lay exactly 24 eggs.

(ii) Use a suitable approximation to calculate the probability that Kamal's hens lay fewer than 20 eggs on any particular day.

Cambridge Paper 6 Q4 J03

12   In a certain city 37% of all shops advertise in the local newspaper.

(i) A random sample of 12 shops is taken. Find the probability that more than 9 advertise in the local newspaper.

(ii) A random sample of 60 shops is taken. Estimate, using a normal approximation, the probability that

(a)   at least 30 advertise in the local newspaper,

(b)   fewer than 39 do **not** advertise in the local newspaper.

13   A manufacturer makes two sizes of elastic bands: large and small. 40% of the bands produced are large bands and 60% are small bands. Assuming that each pack of these elastic bands contains a random selection, calculate the probability that, in a pack of 20 bands, there are

(i) equal numbers of large and small bands,

(ii) more than 17 small bands.

An office pack contains 150 elastic bands.

(iii) Using a suitable approximation, calculate the probability that the number of small bands in the office pack is between 88 and 97 inclusive.

Cambridge Paper 6 Q7 N06

14   On a certain road 20% of the vehicles are trucks, 16% are buses and the remainder are cars.

A random sample of 125 vehicles is taken. Using a suitable approximation, find the probability that more than 73 are cars.

Cambridge Paper 6 Q3(ii) J09

15   A box contains 3 red balloons and 5 green balloons. There are 150 similar boxes in a storeroom. A balloon is taken at random from each box. Using a suitable approximation, find the probability that no more than 100 green balloons are taken.

Summary

Standard normal variable Z

$Z \sim N(0, 1)$ mean = 0, variance = 1, standard deviation = 1

Normal variable X

$X \sim N(\mu, \sigma^2)$ mean = μ, variance = σ^2, standard deviation = σ

To **standardise** X, use $Z = \dfrac{X - \mu}{\sigma}$

To find **probabilities** use the **normal distribution table**

In these illustrations $a > 0, b > 0, a < b$

$P(Z < a) = \Phi(a)$ $P(Z > a) = 1 - \Phi(a)$

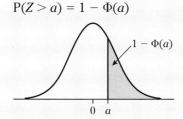

$P(Z > -a) = \Phi(a)$

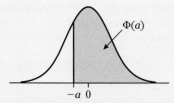

$P(Z < -a) = 1 - \Phi(a)$

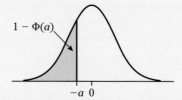

$P(a < Z < b) = \Phi(b) - \Phi(a)$

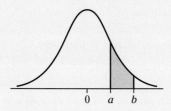

$P(-b < Z < -a) = \Phi(b) - \Phi(a)$

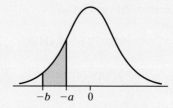

$P(-a < Z < b) = \Phi(b) - (1 - \Phi(a))$

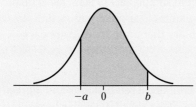

$P(-a < Z < a) = 2\Phi(a) - 1$

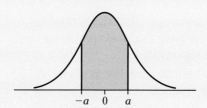

To find z values, read the normal distribution table in reverse, or use the table of **critical values** (page **216**).

If $\Phi(z) = p$, i.e. $P(Z < z) = p$, then $z = \Phi^{-1}(p)$

The normal approximation to the binomial distribution

If $\quad X \sim B(n, p)$ and $np > 5$ and $nq > 5$, where $q = 1 - p$

then $\quad X \sim N(np, npq)$ with $\mu = np$ and $\sigma = \sqrt{npq}$

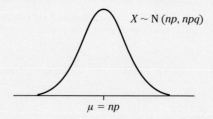

$X \sim N(np, npq)$

$\mu = np$

Continuity corrections

Continuity corrections must be used when calculating binomial probabilities using a normal approximation, for example

$P(X < 32) \rightarrow P(X < 31.5)$ $\qquad$ $P(X \leqslant 32) \rightarrow P(X < 32.5)$

$P(X > 32) \rightarrow P(X > 32.5)$ $\qquad$ $P(X \geqslant 32) \rightarrow P(X > 31.5)$

$P(3 < X < 10) \rightarrow P(3.5 < X < 9.5)$ $\qquad$ $P(3 \leqslant X \leqslant 10) \rightarrow P(2.5 < X < 10.5)$

Mixed Exercise 6

1   The lengths of certain items follow a normal distribution with mean μ mm and standard deviation 6.0 mm. It is known that 4.78% of the items are longer than 82.0 mm.

   (i) Find the value of μ.

   (ii) Find the probability that the length of a randomly selected item is between 63.7 mm and 71.3 mm.

2   The breaking strength of a particular type of paving slab is normally distributed with mean 50 units and standard deviation 4 units.

   (i) Find the probability that a paving slab from a batch of slabs of this type has a breaking strength greater than 59.5 units.

   (ii) 95% of the slabs have a breaking strength between the limits $(50 - b)$ units and $(50 + b)$ units. Find the value of b and hence find these limits.

3   (i) The daily minimum temperature in degrees Celsius (°C) in January in Ottawa is a random variable with distribution $N(-15.1, 62.0)$. Find the probability that a randomly chosen day in January in Ottawa has a minimum temperature above 0 °C.

   (ii) In another city the daily minimum temperature in °C in January is a random variable with distribution $N(\mu, 40.0)$. In this city the probability that a randomly chosen day in January has a minimum temperature above 0 °C is 0.8888. Find the value of μ.

   Cambridge Paper 6 Q3 N08

4   The time spent by customers in a particular supermarket is normally distributed with mean 16.3 minutes and standard deviation 4.2 minutes.

   (i) Find the probability that a customer spends less than 5 minutes in the supermarket.

   (ii) Find the probability that in a random sample of six customers at least one spends more than 25 minutes in the supermarket.

5   In a normal distribution with mean μ and standard deviation σ, $P(X < 16.2) = 0.5$ and $P(X > 18.3) = 0.1049$.

   (i) Write down the value of μ and calculate the value of σ.

   (ii) Find $P(14.5 < X < 15.9)$.

6   Melons are sold in three sizes: small, medium and large. The weights follow a normal distribution with mean 450 grams and standard deviation 120 grams. Melons weighing less than 350 grams are classified as small.

   (i) Find the proportion of melons which are classified as small.

   (ii) The rest of the melons are divided in equal proportions between medium and large. Find the weight above which melons are classified as large.

   Cambridge Paper 6 Q4 J04

7   In a certain country the time taken for a common infection to clear up is normally distributed with mean μ days and standard deviation 2.6 days. 25% of these infections clear up in less than 7 days.

   (i) Find the value of μ.

   In another country the standard deviation of the time taken for the infection to clear up is the same as in part (i), but the mean is 6.5 days. The time taken is normally distributed.

   (ii) Find the probability that, in a randomly chosen case from this country, the infection takes longer than 6.2 days to clear up.

   Cambridge Paper 6 Q4 J08

8   The lengths of fish of a certain type have a normal distribution with mean 38 cm. It is found that 5% of the fish are longer than 50 cm.

   (i) Find the standard deviation.

   (ii) When fish are chosen for sale, those shorter than 30 cm are rejected. Find the proportion of fish rejected.

   (iii) Nine fish are chosen at random. Find the probability that at least one of them is longer than 50 cm.

   Cambridge Paper 6 Q3 J06

9 On any occasion when a particular gymnast performs a certain routine, the probability that she will perform it correctly is 0.65, independently of all other occasions.

(i) Find the probability that she will perform the routine correctly on exactly 5 occasions out of 7.

(ii) On one day she performs the routine 50 times. Use a suitable approximation to estimate the probability that she will perform the routine correctly on fewer than 29 occasions.

(iii) On another day she performs the routine n times. Find the smallest value of n for which the expected number of correct performances is at least 8.

Cambridge Paper 6 Q6 N07

10 The times spent by people visiting a certain dentist are independent and normally distributed with a mean of 8.2 minutes. 79% of people who visit the dentist have visits lasting less than 10 minutes.

(i) Find the standard deviation of the times spent visiting the dentist.

(ii) Find the probability that the time spent visiting this dentist by a randomly chosen person deviates from the mean by more than 1 minute.

(iii) Find the probability that, of 6 randomly chosen people, more than 2 have visits lasting longer than 10 minutes.

(iv) Find the probability that, of 35 randomly chosen people, fewer than 16 have visits lasting less than 8.2 minutes.

Cambridge Paper 63 Q7 N10

11 The length of time a person undergoing a routine operation stays in hospital can be modelled by a normal distribution with mean 7.8 days and standard deviation 2.8 days.

(i) Calculate the proportion of people who spend between 7.8 days and 11.0 days in hospital.

(ii) Calculate the probability that, of 3 people selected at random, exactly 2 spend longer than 11.0 days.

(iii) A health worker plotted a box-and-whisker plot of the times that 100 patients, chosen randomly, stayed in hospital. The results are shown below.

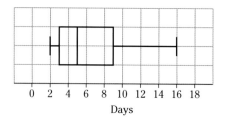

Days

State with a reason whether or not this agrees with the model used in parts (i) and (ii).

Cambridge Paper 6 Q7 J03

12 A shop sells old video tapes, of which 1 in 5 on average are known to be damaged.

(i) A random sample of 15 tapes is taken. Find the probability that at most 2 are damaged.

(ii) Find the smallest value of n if there is a probability of at least 0.85 that a random sample of n tapes contains at least one damaged tape.

(iii) A random sample of 1600 tapes is taken. Use a suitable approximation to find the probability that there are at least 290 damaged tapes.

Cambridge Paper 6 Q7 J04

13 (i) In a spot check of the speeds, $x\,\text{km h}^{-1}$, of 30 cars on a motorway, the data were summarised by $\sum(x - 110) = -47.2$ and $\sum(x - 110)^2 = 5460$. Calculate the mean and standard deviation of these speeds.

(ii) On another day the mean speed of cars on the motorway was found to be $107.6\,\text{km h}^{-1}$ and the standard deviation was $13.8\,\text{km h}^{-1}$. Assuming these speeds follow a normal distribution and that the speed limit is $110\,\text{km h}^{-1}$, find what proportion of cars exceed the speed limit.

Cambridge Paper 6 Q4 J02

Normal distribution tables

The normal distribution function

If Z has a normal distribution with mean 0 and variance 1 then, for each value of z, the table gives the value of $\Phi(z)$, where

$$\Phi(z) = P(Z \leqslant z).$$

For negative values of z use $\Phi(-z) = 1 - \Phi(z)$.

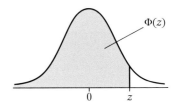

z	0	1	2	3	4	5	6	7	8	9	1	2	3	4	5	6	7	8	9
															ADD				
0.0	0.5000	0.5040	0.5080	0.5120	0.5160	0.5199	0.5239	0.5279	0.5319	0.5359	4	8	12	16	20	24	28	32	36
0.1	0.5398	0.5438	0.5478	0.5517	0.5557	0.5596	0.5636	0.5675	0.5714	0.5753	4	8	12	16	20	24	28	32	36
0.2	0.5793	0.5832	0.5871	0.5910	0.5948	0.5987	0.6026	0.6064	0.6103	0.6141	4	8	12	15	19	23	27	31	35
0.3	0.6179	0.6217	0.6255	0.6293	0.6331	0.6368	0.6406	0.6443	0.6480	0.6517	4	7	11	15	19	22	26	30	34
0.4	0.6554	0.6591	0.6628	0.6664	0.6700	0.6736	0.6772	0.6808	0.6844	0.6879	4	7	11	14	18	22	25	29	32
0.5	0.6915	0.6950	0.6985	0.7019	0.7054	0.7088	0.7123	0.7157	0.7190	0.7224	3	7	10	14	17	20	24	27	31
0.6	0.7257	0.7291	0.7324	0.7357	0.7389	0.7422	0.7454	0.7486	0.7517	0.7549	3	7	10	13	16	19	23	26	29
0.7	0.7580	0.7611	0.7642	0.7673	0.7704	0.7734	0.7764	0.7794	0.7823	0.7852	3	6	9	12	15	18	21	24	27
0.8	0.7881	0.7910	0.7939	0.7967	0.7995	0.8023	0.8051	0.8078	0.8106	0.8133	3	5	8	11	14	16	19	22	25
0.9	0.8159	0.8186	0.8212	0.8238	0.8264	0.8289	0.8315	0.8340	0.8365	0.8389	3	5	8	10	13	15	18	20	23
1.0	0.8413	0.8438	0.8461	0.8485	0.8508	0.8531	0.8554	0.8577	0.8599	0.8621	2	5	7	9	12	14	16	19	21
1.1	0.8643	0.8665	0.8686	0.8708	0.8729	0.8749	0.8770	0.8790	0.8810	0.8830	2	4	6	8	10	12	14	16	18
1.2	0.8849	0.8869	0.8888	0.8907	0.8925	0.8944	0.8962	0.8980	0.8997	0.9015	2	4	6	7	9	11	13	15	17
1.3	0.9032	0.9049	0.9066	0.9082	0.9099	0.9115	0.9131	0.9147	0.9162	0.9177	2	3	5	6	8	10	11	13	14
1.4	0.9192	0.9207	0.9222	0.9236	0.9251	0.9265	0.9279	0.9292	0.9306	0.9319	1	3	4	6	7	8	10	11	13
1.5	0.9332	0.9345	0.9357	0.9370	0.9382	0.9394	0.9406	0.9418	0.9429	0.9441	1	2	4	5	6	7	8	10	11
1.6	0.9452	0.9463	0.9474	0.9484	0.9495	0.9505	0.9515	0.9525	0.9535	0.9545	1	2	3	4	5	6	7	8	9
1.7	0.9554	0.9564	0.9573	0.9582	0.9591	0.9599	0.9608	0.9616	0.9625	0.9633	1	2	3	4	4	5	6	7	8
1.8	0.9641	0.9649	0.9656	0.9664	0.9671	0.9678	0.9686	0.9693	0.9699	0.9706	1	1	2	3	4	4	5	6	6
1.9	0.9713	0.9719	0.9726	0.9732	0.9738	0.9744	0.9750	0.9756	0.9761	0.9767	1	1	2	2	3	4	4	5	5
2.0	0.9772	0.9778	0.9783	0.9788	0.9793	0.9798	0.9803	0.9808	0.9812	0.9817	0	1	1	2	2	3	3	4	4
2.1	0.9821	0.9826	0.9830	0.9834	0.9838	0.9842	0.9846	0.9850	0.9854	0.9857	0	1	1	2	2	2	3	3	4
2.2	0.9861	0.9864	0.9868	0.9871	0.9875	0.9878	0.9881	0.9884	0.9887	0.9890	0	1	1	1	2	2	2	3	3
2.3	0.9893	0.9896	0.9898	0.9901	0.9904	0.9906	0.9909	0.9911	0.9913	0.9916	0	1	1	1	1	2	2	2	2
2.4	0.9918	0.9920	0.9922	0.9925	0.9927	0.9929	0.9931	0.9932	0.9934	0.9936	0	0	1	1	1	1	1	2	2
2.5	0.9938	0.9940	0.9941	0.9943	0.9945	0.9946	0.9948	0.9949	0.9951	0.9952	0	0	0	1	1	1	1	1	1
2.6	0.9953	0.9955	0.9956	0.9957	0.9959	0.9960	0.9961	0.9962	0.9963	0.9964	0	0	0	0	1	1	1	1	1
2.7	0.9965	0.9966	0.9967	0.9968	0.9969	0.9970	0.9971	0.9972	0.9973	0.9974	0	0	0	0	0	1	1	1	1
2.8	0.9974	0.9975	0.9976	0.9977	0.9977	0.9978	0.9979	0.9979	0.9980	0.9981	0	0	0	0	0	0	0	1	1
2.9	0.9981	0.9982	0.9982	0.9983	0.9984	0.9984	0.9985	0.9985	0.9986	0.9986	0	0	0	0	0	0	0	0	0

Critical values for the normal distribution

If Z has a normal distribution with mean 0 and variance 1 then, for each value of p, the table gives the value of z such that

$$P(Z \leqslant z) = p.$$

p	0.75	0.90	0.95	0.975	0.99	0.995	0.9975	0.999	0.9995
z	0.674	1.282	1.645	1.960	2.326	2.576	2.807	3.090	3.291

List of formulae provided in the exam

Probability and statistics

Summary statistics

For ungrouped data:

$$\bar{x} = \frac{\Sigma x}{n}, \qquad \text{standard deviation} = \sqrt{\frac{\Sigma(x - \bar{x})^2}{n}} = \sqrt{\frac{\Sigma x^2}{n} - \bar{x}^2}$$

For grouped data:

$$\bar{x} = \frac{\Sigma xf}{\Sigma f}, \qquad \text{standard deviation} = \sqrt{\frac{\Sigma(x - \bar{x})^2 f}{\Sigma f}} = \sqrt{\frac{\Sigma x^2 f}{\Sigma f} - \bar{x}^2}$$

Discrete random variables

$$E(X) = \Sigma xp$$
$$\text{Var}(X) = \Sigma x^2 p - \{E(X)\}^2$$

For the binomial distribution $B(n, p)$:

$$p_r = \binom{n}{r} p^r (1 - p)^{n - r}, \qquad \mu = np, \qquad \sigma^2 = np(1 - p)$$

For the Poisson distribution $Po(a)$:

$$p_r = e^{-a} \frac{a^r}{r!}, \qquad \mu = a, \qquad \sigma^2 = a$$

Continuous random variables

$$E(X) = \int xf(x)\,dx$$
$$\text{Var}(X) = \int x^2 f(x)\,dx - \{E(X)\}^2$$

Sampling and testing

Unbiased estimators:

$$\bar{x} = \frac{\Sigma x}{n}, \qquad s^2 = \frac{1}{n - 1}\left(\Sigma x^2 - \frac{(\Sigma x)^2}{n}\right)$$

Central Limit Theorem:

$$\bar{X} \sim N\left(\mu, \frac{\sigma^2}{n}\right)$$

Approximate distribution of sample proportion:

$$N\left(p, \frac{p(1 - p)}{n}\right)$$

Sample exam papers

*Answer **all** the questions.*

Give non-exact numerical answers correct to 3 significant figures, or 1 decimal place in the case of angles in degrees, unless a different level of accuracy is specified in the question. The use of an electronic calculator is expected, where appropriate. You are reminded of the need for clear presentation in your answers.

The number of marks is given in brackets [ ] at the end of each question or part question.

The total number of marks for this paper is 50.

Questions carrying smaller numbers of marks are printed earlier in the paper, and questions carrying larger numbers of marks later in the paper.

Note: The number of marks for each question reflects the amount of working required in the answer.

Paper 1

Q1 In a store, 30% of loaves come from bakery A and the rest come from bakery B. 20% of the loaves from bakery A are brown and the rest are white. 60% of the loaves from bakery B are brown and the rest are white. A loaf from the store is chosen at random.

   (i) Find the probability that the loaf is a white loaf from bakery A. [2]

   (ii) Given that the loaf is white, find the probability that it is from bakery A. [3]

Q2 (i) The life, in hours, of a certain kind of battery has the distribution $N(5.5, 1.1^2)$. Find the probability that a randomly chosen battery of this kind has a life longer than 6 hours. [3]

   (ii) The life, in hours, of another kind of battery has the distribution $N(7.2, \sigma^2)$. The probability that a randomly chosen battery of this kind has a life longer than 6 hours is 0.923. Find the value of σ. [3]

Q3 A bag contains 2 red discs and 3 blue discs. Three discs are taken at random from the bag, without replacement. The random variable X is the number of red discs that are taken.

   (i) Draw up a table for the probability distribution of X. [3]

   (ii) Calculate $E(X)$ and $Var(X)$. [4]

Q4 A class consists of 8 men and 5 women. A committee of 7 people is chosen at random from the class.

   (i) Find the probability that 4 men and 3 women are chosen. [3]

In fact 4 men and 3 women are chosen and all 7 committee members sit in a row. In how many different orders can they sit if

   (ii) no two men sit next to each other, [2]

   (iii) Rajiv and John sit next to each other? [2]

Q5 On average, 40% of packets of cereal contain a token. In order to claim a free gift, 7 tokens are needed.

   (i) Suzy buys one packet of cereal each week for 10 weeks. Find the probability that

     (a) Suzy will be able to claim a free gift at some time during the 10 weeks, [3]

     (b) Suzy will be able to claim a free gift in the 10th week but not before. [3]

(ii) Charlie buys one packet of cereal each week for 50 weeks. Use an appropriate approximating distribution to find the probability that he will be able to claim at least two free gifts during the 50 weeks. [5]

Q6 40 people were asked how long they spent travelling to work. The results are summarised below. Times are given to the nearest minute.

Time (t minutes)	$20 \leqslant t \leqslant 22$	$23 \leqslant t \leqslant 25$	$26 \leqslant t \leqslant 30$	$31 \leqslant t \leqslant 40$
Frequency	1	13	20	6

(i) (a) Calculate estimates of the mean and standard deviation of the time spent travelling to work. [6]

(b) Explain why your answers are only estimates. [1]

(ii) Draw, on graph paper, a histogram to illustrate the data. [5]

(iii) State which class contains the median value of t. Explain your answer. [2]

Paper 2

Q1 Demos measured the lengths, x millimetres, of 50 worms. His results are summarised below.

$$\sum(x - 20) = -24.5 \qquad \sum(x - 20)^2 = 15.24$$

Calculate the mean and standard deviation of the lengths of these worms. [4]

Q2 The masses, in grams, of apples have the distribution $N(\mu, \sigma^2)$. Apples with masses less than $75\,g$ are classified as 'Small'. Apples with masses greater than $110\,g$ are classified as 'Large'. Given that 25% of apples are small and 20% of apples are large, calculate μ and σ. [6]

Q3 Jasmine has 5 identical shoe boxes, of which 2 contain a pair of red shoes and 3 contain a pair of blue shoes. Jasmine wants 2 pairs of red shoes, but she has forgotten which boxes they are in. She opens boxes at random until she finds both pairs of red shoes.

Let X be the number of boxes that Jasmine opens.

(i) Copy and complete the table showing the probability distribution of X.

x	2	3	4	5
$P(X = x)$				$\frac{2}{5}$

[4]

(ii) Find $E(X)$. [2]

Q4 The examination marks of 18 students were displayed in a stem-and-leaf diagram and stored on a computer. Unfortunately, the data became corrupted and some figures were lost. The diagram shows the data that were left.

```
2 | 3
3 | 2 6
4 | 1 3 5
5 | 3 p q 6       Key: 3 | 6 means 36 marks
```

(i) Find the lower quartile. [1]

(ii) Given that the median is 55, find all the possible values of p and q. [2]

(iii) Given that the upper quartile is 61, copy and complete a possible stem-and-leaf diagram. [2]

(iv) State one advantage of a stem-and-leaf diagram over a box-and-whisker plot. [1]

(v) State one advantage of a box-and-whisker plot over a stem-and-leaf diagram. [1]

Q5 On average, 5% of the dolls made at a factory are faulty.

(i) A random sample of 30 dolls is checked. Find the probability that at least 4 of these dolls are faulty. [3]

(ii) A random sample of 220 dolls is checked. Use a suitable approximation to find the probability that more than 13 of these dolls are faulty. [5]

Q6 An examination consists of a Pure Mathematics section containing 5 questions, A, B, C, D and E, and a Statistics section containing 3 questions, X, Y and Z.

(i) The questions are arranged in a random order within each section.

    (a) How many different arrangements are possible if questions E and X are next to each other? [2]

    (b) How many different arrangements are possible if there are at least 3 questions between questions E and X? [4]

(ii) Candidates have to choose 5 questions, including at least 2 Pure Mathematics questions and at least 1 Statistics question. The order in which they choose the questions does not matter. How many selections of 5 questions are possible? [3]

Q7 Jhoti plays one game of tennis against each of Ann, Beryl and Chloe. The independent probabilities of Jhoti winning the games are 0.6, 0.3 and 0.8 respectively.

(i) Draw a tree diagram to show this information. [2]

Find the probability that

(ii) Jhoti loses at least 1 game, [2]

(iii) Jhoti wins exactly 1 game, [3]

(iv) Jhoti does not win against Ann, given that she wins exactly one game. [3]

S1 Answers

The University of Cambridge Local Examinations Syndicate bears no responsibility for the example answers to questions taken from its past question papers which are contained in this publication.

Chapter 1

Exercise 1a

1

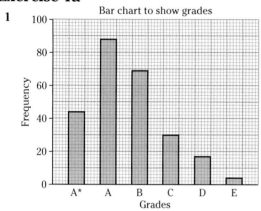

Bar chart to show grades

2 $208°, 46°, 38°, 36°, 32°$

3 (i) $150°, 30°, 75°, 105°$

    (ii) Pie chart to show how pupils travelled to school

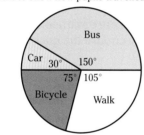

4

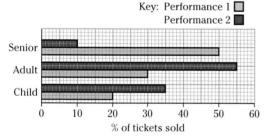

    Other diagrams possible.

5 (i)

Bar chart to show golf club members

(ii) Pie chart chart to show golf club members

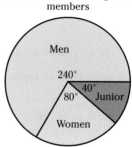

6 68 000

Exercise 1b

1 (i) 18   (ii) 66

2 (i) **Key:** 5 | 2 means 52 kg

5	2 7 7 9
6	1 3 4 4 5 7 7 8 8 8
7	0 1 1 2 3 4 4 5 6 6 7 9 9
8	1 3 6

    (ii) 68

3 (i) **Key:** 5 | 3 means 5.3 cm

3	9
4	
5	3 4 5 5
6	1 1 5 7 8
7	0 0 1 3 4 5 6 6 8 9
8	0 1 2 2 4 8
9	2 6
10	0 1

    (ii) 47%

4 **Key:** 7 | 3 means 7.3 hours

0	0 2 6 8
1	6
2	4 6
3	
4	3 8
5	6 6 8
6	1 2 8
7	0 3 5 5 6 8
8	3 4
9	7 8
10	4
11	1 3 6
12	5 9

5 (i) **Key:** 201 | 5 means 2.015 litres

198	2 5 6 8
199	0
200	1 3 4 8
201	1 5 7 7 8 9 9
202	2 3 4
203	3

(ii) 25%

6 (i) **Key:** 4 | 6 | 9 means 64 Before exercise
                            69 After exercise

Before exercise		After exercise
8	4	
7 3 1 1 0	5	
9 9 6 6 4	6	9
9 5 3 3 0 0	7	0 5 5 7 7
1	8	0 0 1 4 4 6
3 3 3 3 1 0 0	9	5 6 7
5 5	10	4 4 4 6 8 9
1 1 0	11	7
	12	5
	13	0 0 1 7 7
	14	3 5

(ii) Pulse rate much faster after exercise

7 (i) **Key:** 1 | 3 | 4 means age 31 in School A
                            age 34 in School B

School A		School B
9 8 7 5 3 3	2	3 5 9
9 9 9 7 7 7 4 3 3 1 1	3	4 6 6 8 8
8 8 8 8 6 6 5 5 5 0 0	4	0 1 2 2 3 4 5 5 6 7 7 9
9 4 4 3 3 1 1	5	0 0 2 2 4 4 6 6 6 7 8 8 9 9 9
1	6	0

(ii) Older teachers in School B

Exercise 1c

1 (i) 5

(ii) 0.4, 1.2, 1.4, 1, 0.4

(iii)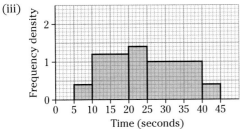

(iv) $20 \leqslant t < 25$

2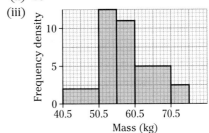

(ii) $60 \leqslant t < 80$

3 (i) 40.5, 50.5

(ii) 10

(iii)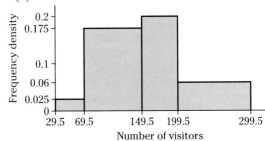

4 (i) 29.5, 69.5

(ii)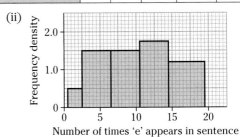

5 (i)

Number of times 'e' appears	1–2	3–6	7–10	11–14	15–19
Frequency	1	6	6	7	6

(ii)

6 (i) −0.5, 1.5

(ii)

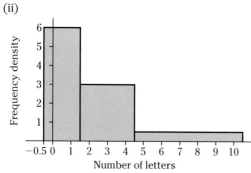

7 (i)

Mass (g)	Frequency	f.d.
85–89	4	0.8
90–94	6	1.2
95–99	7	1.4
100–104	13	2.6
105–109	10	2
110–114	5	1
115–119	5	1

(ii) 89.5

(iii)

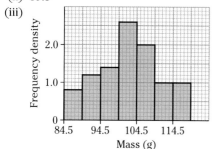

Modal class is 100–104

(iv) **Key:** 10 | 3 means 103

```
 8 | 6 6 7 8
 9 | 2 2 2 2 3 3
 9 | 5 6 6 7 8 9 9
10 | 0 0 0 1 1 1 1 1 2 2 3 3 4
10 | 5 5 5 6 6 7 7 8 8 9
11 | 0 1 3 3 4
11 | 6 6 7 8 8
```

mode = 101 g

8 (i) Frequencies are 20, 24, 24, 16, 12, 10, 6
  (ii) 112
9 (i) 49.5, 53.5, width 4   (ii) 20, 26, 12
  (iii) 88
10 (i) 0.5, 10.5, width 10   (ii) 60
  (iii) 380   (iv) 100

Exercise 1d

1 (i) 9.7   (ii) 154.8   (iii) 50.875
  (iv) $1775\frac{5}{7}$ (v) 0.908 (3 s.f.)
  (vi) 4   (vii) 29.54

2 35.05
3 43.35
4 (i) 33   (ii) 44.875
5 14.96 m (2 d.p.)
6 (i) 28.1 seconds   (ii) 30–31
7 60.9
8 $a = 12$
9 (i) 16   (ii) 40
10 $x = 15, y = 7$
11 19
12 4
13 8
14 7

Exercise 1e

1 (a) 5, 2   (b) 8.5, 1.80   (c) 18.8, 6.46
  (d) $10\frac{5}{6}$, 4.10   (e) 3.42, 1.91   (f) 205, 3.16
2 14.39, 1.08
3 11.52, 0.827
4 69.3, 1.7
5 6.8, 1.11
6 (i) 10   (ii) 3.42
7 (i) 0.6   (ii) 0.24
8 207.62, 77.93
9 115.8, 7.58
10 12.4, 3.87
11 (i) −0.5, 2.5   (ii) 5.29, 3.58
12 (i) Frequencies 5, 18, 22, 28, 22, 18, 5
  (ii) The histogram is symmetrical, with a line of symmetry through 111, so an estimate of the mean is 111. This may not be the true mean as the mid-interval values have been taken to represent the intervals.
  (iii) 3.71
13 (i) 63.87   (ii) 29.47   (iii) 133, 144

Exercise 1f

1 1.014, 0.0102
2 (i) 29   (ii) 2.429   (iii) 5.9
3 (i) 5.84   (ii) 203.7
4 5.099
5 (i) 2.236   (ii) 4.33
6

	n	$\sum x$	$\sum x^2$	$\bar{x}$	s.d.
(i)	63	7623	924 800	**121**	**6.194**
(ii)	**14**	152.6	**1703.8**	10.9	1.7
(iii)	52	**1716**	57 300	33	**3.595**
(iv)	18	**1026**	**58 770**	57	4

7 (i) 49.85   (ii) 0.5275
8 3.838
9 3.742
10 (i) 5, 2.739   (ii) 5, 11
11 $a = 4, b = 6$

12 (i) 1800   (iii) 17.436
13 (i) 4.6, 2   (ii) 4.56, 2.043
14 2.3, 0.871
15 51.235, 0.927
16 (i) more than 25 kg   (ii) 25.34
      (iii) 0.974
17 (i) 15.83   (ii) 0
18 (i) 2500   (ii) 500   (iii) −200
      (iv) 63 400   (v) 3400   (vi) 1300

Exercise 1g

1 (i) 61   (ii) 52
    (iii) 73   (iv) 21
2 (i) (a) 10   (b) 207   (c) 2104.5
        (d) 0.595   (e) −1.5
    (ii) (a) 12   (b) 22   (c) 765.5
        (d) 0.38   (e) 3.35
3 (i) 4   (ii) 2, 5   (iii) 3
4 (i) 4.6, 3.5   (ii) 0.138, 0.012
5 (i) 8   (ii) 7   (iii) 7.36
6 (i) 4   (ii) 3   (iii) 1.228
7 (i) 2   (ii) 2
8 (i) 2   (ii) 3
    (iii)

Times absent	0	1	2	3	4	5	6	7
Frequency	5	6	9	3	4	1	3	1

    (iv) 2.47   (v) 1.936
9 (i) 24, 16   (ii) 5

Exercise 1h

When reading from cumulative frequency graphs, answers will depend on your graph.
1 (i) 22   (ii) 26   (iii) 23
    (iv) 25 mins   (v) 9 mins
2 (i)

Mass (x grams)	<50	<54	<58	<62	<66	<70	<74
Cumulative frequency	6	10	20	44	64	76	80

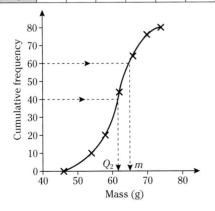

(ii) 62 g
(iii) 65
3 (i)

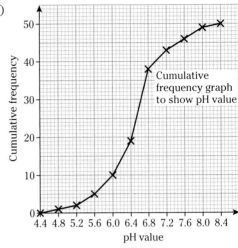

Cumulative frequency graph to show pH value

(ii) 82%
(iii) 6.5
(iv)

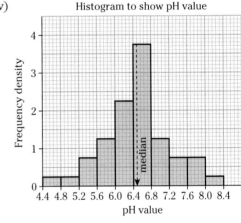

Histogram to show pH value

4 (i) 688, 11

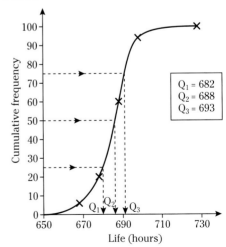

$Q_1 = 682$
$Q_2 = 688$
$Q_3 = 693$

(ii) 686.8, 11.58
5 (i) −0.5, 4.5, 5

(ii)

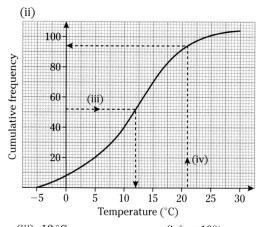

(iii) 12 °C   (iv) ≈10%

6 (i)

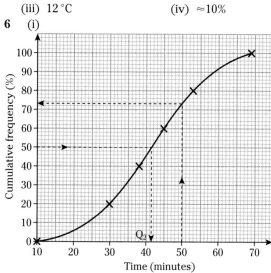

(ii) 41.5 mins   (iii) 73%

7 (i) Frequencies: 14, 31, 25, 18, 12

(ii) $25 \leqslant x < 30$

(iii)

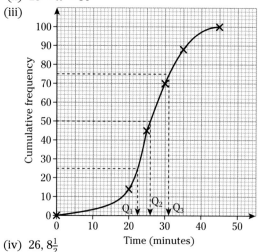

(iv) 26, $8\frac{1}{2}$

8 (a) (i)

Distance (x km)	<4	<10	<20	<35	<60
Cumulative frequency	5	15	54	149	200

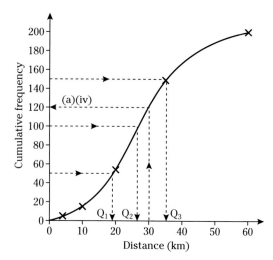

(ii) 27.5 km

(iii) 17 km

(iv) 40%

(b) (i)   f.d. 1.25, 1.67, 3.9, 6.33, 2.04

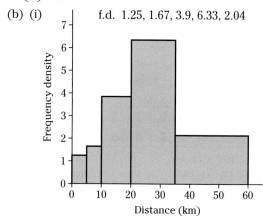

(ii) 28.5, 13.0

9 (i)

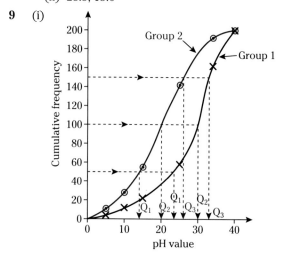

(ii) Group 1: 20, 12; Group 2: 30, 11

(iii) Group 2 sent more texts

10   (i)

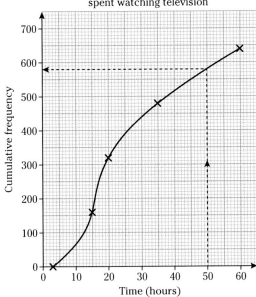

Cumulative frequency graph to show time spent watching television

(ii) 60 [40–70 acceptable, depending on graph].

11

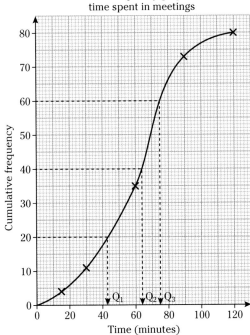

Cumulative frequency graph to show time spent in meetings

64 mins, 32 mins (depends on graph)

Exercise 1i

1   (i) 26, 38   (ii) 32   (iii) 32
       (iv) 12   (v) Symmetrical

2

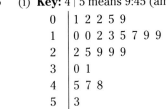

3   (i) Negative skew

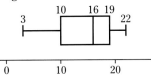

(ii) Symmetrical

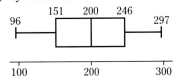

(iii) Positive skew

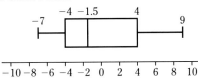

4   (i) 6.6; 5.7, 7.8
       (ii)

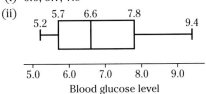

Blood glucose level

5   (i) **Key:** 4 | 5 means 9:45 (am)

0	1 2 2 5 9
1	0 0 2 3 5 7 9 9
2	2 5 9 9 9
3	0 1
4	5 7 8
5	3

(ii) 9:19 am   (iii) 9:10 am, 29.5 minutes after 9
(iv)

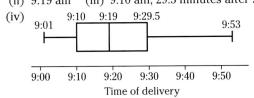

Time of delivery

6

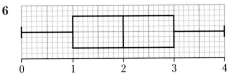

7   (i) 1, 8
       (ii)

x	≤1	≤2	≤3	≤4	≤5	≤6	≤7	≤8
Cumulative frequency	3	8	14	20	28	42	54	63

(iii) 6, 4, 7
(iv)

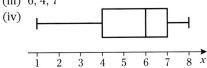

8 (i) 35, 22, 51, 29

(ii)
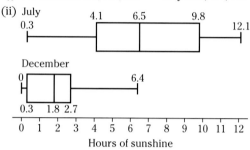
Length of line (mm)

(iii) Boundary for outliers 94.5; outlier 97

9 (i) December: 1.8; 0.3, 2.7   July: 6.5; 4.1, 9.8

(ii)
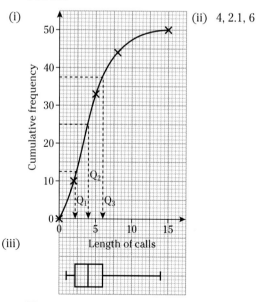
Hours of sunshine

(iii) The median number of hours of sunshine in July is greater than the maximum number in December. The mode for December is 0.0 hours and the mode for July is 6.6 hours. There is much greater variation in the number of hours of sunshine in July than in December. In July the hours of sunshine are evenly spread throughout the range, whereas in December there is a long tail to the right with the bulk of values being under 3 hours.

10 (i)

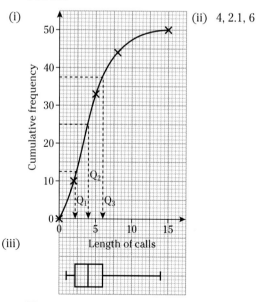

(ii) 4, 2.1, 6

(iii)

There were a few unusually long calls.

11 Surgery B has shorter waiting times on average than Surgery A. Surgery B has a greater variability (spread) of waiting times than Surgery A.

12 (i) **Key:** 4 | 1 means 41 km

Distance travelled (km)

4	1 2 3 4 4 6 7 7 8 8
5	0 2 2 2 3 4 6 7 8 8
6	0 2 3 3 6 6 7 7 8
7	0 0 2 2 4 4 6 7 8 8 8
8	0 1 2 5 5 6 6 7
9	3 3 4

(ii) 66, 52, 78

(iii)
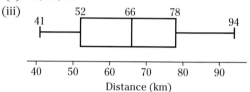
Distance (km)

(iv) (a) The stem-and-leaf diagram retains all the original data.

(b) The box-and-whisker plot shows that the distribution is approximately symmetrical with the central 50% lying between 52 and 78 km.

13 (i) New: 450, 397, 426;
Standard: 368.5, 353, 383

(ii)
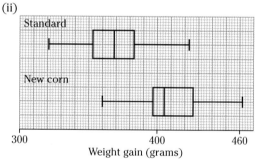
Weight gain (grams)

(iii) The weight gain was much greater on average in the chicks fed the new strain of corn. The variation in weight gain was similar in the two groups.

14 (i) 55, 44, 74   (ii) 89, 71, 107

(iii)
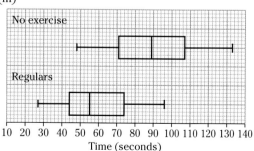
Time (seconds)

(iv) The breathing rate took much less time to return to normal in the regular gym users than in those who did not exercise regularly.

Mixed Exercise 1

1 *The intervals could be written 42–45, 46–49, 50–53, 54–57, 58–61 or* $42 \leqslant w \leqslant 45$, $46 \leqslant w \leqslant 49$, $50 \leqslant w \leqslant 53$, $54 \leqslant w \leqslant 57$, $58 \leqslant w \leqslant 61$

Weight (nearest kg)	Frequency
41.5–45.5	4
45.5–49.5	7
49.5–53.5	10
53.5–57.5	5
57.5–61.5	4

2 38.4 mm, 4.57 mm

3 (i) As the vertical axis does not start at 0, the vertical scale is distorted, making the increase appear larger than it is.

   (ii) (a) **Key:** 8 | 6 means 86

Daily ticket sales

```
3 | 4 5
4 | 1 4 5
5 | 0 2
6 | 2
7 | 3 3 9
8 | 3 4 4 5 5 6 6 7 9
9 | 1
```

     (b) 79

4 (i) Two modes: 3.05, 3.45   (ii) 3.25

   (iii) 3.321   (iv) 50%

5 (i) 1850   (iii) 74.72, 8.234

6 *Other answers are possible, for example comparative bar charts*

Pie charts to show ages of car drivers

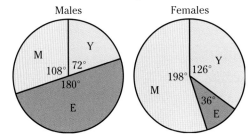

7 (i)

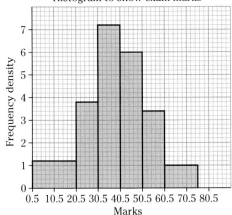

8 (ii) 38.74, 14.59

   (i) 61, 73, 83.6

   (ii) 181, 21, 42.0

   (iii) (a) median (not affected by outlier)

       (b) interquartile range (not affected unduly by outliers)

9 (i) **Key:** 13 | 2 means 132 beats/min

Pulse rate (beats per minute)

```
10 | 4 4 9
11 | 5 7
12 | 0 4 5
13 | 2 4
14 | 2 5
15 | 8
16 | 0 2
```

   (ii) 125, 115, 145

   (iii)

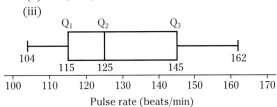

10 0.850, 0.9775

11 (i)

Mass (kg)	<39.5	<44.5	<49.5	<54.5	<59.5	<64.5	<69.5	<74.5
Cumulative frequency	0	3	5	12	30	48	51	52

   (ii) 21

   (iii) 14

   (iv) 62

   (v) 58.4 kg

   (vi) 7.2 kg

12 33.75 mins, 2.3 mins

13 (i) B

   (ii) A

   (iii) C

14 (i) 78, 72, 88

   (ii) Country P

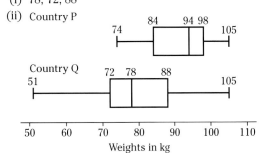

   (iii) Males are heavier in country P; weights are more variable in country Q

15 45

16 (a) 180.58 cm, 6.85 cm

(b) (i)

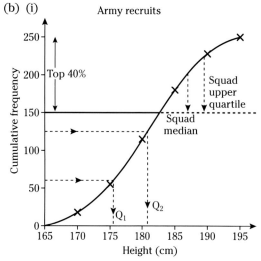

Army recruits

(ii) 180.6 cm, 176 cm
(iii) 187 cm
(iv) 189.5 cm

17 (i) 9 secs   Lap times of athletes

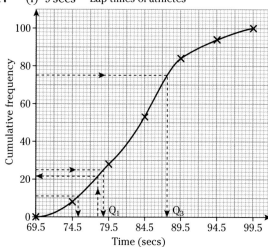

(ii) 22, 75.5 secs

18 (i) **Key:** 7 | 13 | 2 means
13.7 minutes for 16-yr-olds
13.2 minutes for 9-yr-olds

16-yr-olds		9-yr-olds
7 4	11	
9 8	12	
7 0	13	0 2 7
8	14	2 4
	15	0 1 9
5	16	0 1 4 7

(ii) 15.6 mins

19 Medians A: 2.0, B: 3.8;
Country B has heavier babies
Interquartile ranges A: 0.9, B: 2.3;
Country B has greater spread of weights

20 (i) $a = 494, b = 46$

(ii)

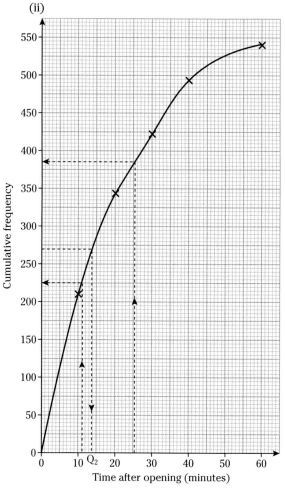

(iii) 14 minutes (13.5 to 14.6 is acceptable)
(iv) $m = 18.204, s = 14.168...$
(v) 160

Chapter 2

Exercise 2a

Section A

1 (i) 5040   (ii) 720
  (iii) 60   (iv) 140
2 (i) 4032   (ii) 252

3 (i) 5!   (ii) $\dfrac{5!}{2!}$ (iii) $\dfrac{21!}{17!}$

  (iv) $\dfrac{n!}{(n-3)!}$ (v) $\dfrac{12!}{8!} \times \dfrac{2!}{5!}$

Section B

1 (i) 720   (ii) 5040   (iii) 362 880
2 (i) 720   (ii) 120   (iii) 40 320
3 (i) 4! = 24   (ii) $4^4 = 256$
4 (i) 3! = 6   (ii) 12
5 (i) 2520   (ii) 907 200   (iii) 19 958 400
6 360
7 (i) 3360   (ii) 420   (iii) 360
  (iv) 2520   (v) 240

8 (i) 60   (ii) 20   (iii) 30
9 243
10 (i) 10 000   (ii) 59 049
11 2880
12 (i) 10 080   (ii) 1440
13 (i) 2 903 040   (ii) 241 920
14 3840
15 5040
16 (i) 483 840   (ii) 1 209 600
17 98
18 (i) 40 320   (ii) 2880
19 (i) 50   (ii) 18
20 (i) 90 720   (ii) 120

Exercise 2b

1 (i) 60 480   (ii) 30
     (iii) 720   (iv) 362 880
2 6840
3 11 880
4 210
5 12
6 (i) 1200   (ii) 3125
7 151 200
8 (i) 60   (ii) 325
9 (i) 151 200   (ii) 30 240

Exercise 2c

1 (i) 330   (ii) 35   (iii) 56
     (iv) 45   (v) 84   (vi) 84
     (vii) 495   (viii) 495
2 792
3 (i) 126   (ii) 56   (iii) 105
     (iv) 15
4 1050
5 (i) 7920   (ii) 5544   (iii) 7182
6 10
7 168 168
8 25 200
9 (i) 15   (ii) 75
10 26
11 7
12 (i) 384   (ii) 86   (iii) 2590
13 (ii) 9

Mixed Exercise 2

1 (i) (a) 9 979 200   (b) 181 440
     (ii) 15
2 (i) 33 033 000   (ii) 86 400   (iii) 288
3 (i) 362 880   (ii) 282 240   (iii) 504
     (iv) 168   (v) 476
4 (i) 512   (ii) 151 200
5 (a) (i) 60   (ii) 216
     (b) (i) 1316   (ii) 517

6 (i) 2520   (ii) 360   (iii) 1440
7 91
8 (i) 4.94×10^{11} (ii) 79 833 600   (iii) 21
9 (ii) 162   (iii) $162^4 = 688\,747\,536$

Chapter 3

Exercise 3a

1 (i) $\frac{1}{2}$ (ii) 1   (iii) $\frac{2}{3}$ (iv) $\frac{1}{2}$
     (v) $\frac{5}{6}$
2 (i) $\frac{3}{8}$ (ii) $\frac{5}{8}$ (iii) 0   (iv) $\frac{4}{5}$
3 (i) $\frac{3}{10}$ (ii) $\frac{3}{4}$
4 (i) (a) $\frac{1}{13}$ (b) $\frac{1}{2}$ (c) $\frac{3}{52}$
     (ii) $\frac{7}{25}$
5 (i) 0.4   (ii) 0.5   (iii) 0.25
6 (i) $a = 18, b = 14$
     (ii) (a) 0.29   (b) 0.1   (c) 0.75
7 $\frac{4}{15}$
8 $\frac{4}{15}$
9 (i) $\frac{2}{7}$ (ii) $\frac{3}{7}$
10 (i) 0.26   (ii) 0.06   (iii) 0.46   (iv) 0
11 (i) $\frac{1}{4}$ (ii) $\frac{3}{4}$ (iii) $\frac{3}{8}$
12 (i) $\frac{1}{2}$ (ii) $\frac{3}{4}$
13 (i) $\frac{1}{18}$ (ii) $\frac{1}{6}$ (iii) $\frac{1}{6}$ (iv) $\frac{1}{3}$
14 (i) (a) $\frac{1}{36}$ (b) $\frac{1}{12}$ (c) 0   (d) 0
     (ii) 6, 12
15 (i) HHH, HHT, HTH, THH, HTT, THT, TTH, TTT
     (ii) $\frac{3}{8}$

Exercise 3b

1 (i) 362 880   (ii) $\frac{1}{72}$
2 (i) 8 709 120   (ii) $\frac{9}{30}$
3 $\frac{1}{126}$
4 (i) $\frac{60}{143}$ (ii) $\frac{85}{1001}$
5 (i) $\frac{3}{5}$ (ii) $\frac{4}{5}$
6 (i) $\frac{28}{153}$ (ii) $\frac{5}{17}$ (iii) $\frac{80}{153}$
7 (i) $\frac{2}{11}$ (ii) $\frac{9}{11}$
8 $\frac{37}{42}$
9 (i) $\frac{7}{1938}$ (ii) $\frac{25}{1292}$ (iii) $\frac{455}{1292}$ (iv) $\frac{1001}{7752}$
10 (i) 84   (ii) $\frac{1}{3}$
11 (i) (1, 2, 3), (1, 3, 2), (2, 1, 3), (2, 3, 1), (3, 1, 2),
        (3, 2, 1)
     (ii) $\frac{3}{32}$
12 (i) $\frac{20}{91}$ (ii) $\frac{24}{91}$ (iii) $\frac{12}{65}$
13 (i) 64   (ii) $\frac{21}{32}$
14 $0.0468\ldots = 4.7\%$ (2 s.f.) $< 5\%$

Exercise 3c

1 (i) $\frac{1}{2}$ (ii) $\frac{1}{2}$ (iii) $\frac{5}{6}$ (iv) $\frac{1}{6}$

2 (i) $\frac{11}{30}$ (ii) $\frac{9}{30}$

3 (i) $\frac{4}{17}$ (ii) $\frac{4}{51}$ (iii) $\frac{5}{17}$ (iv) $\frac{5}{17}$
    (v) 0

4 (i) 0.41 (ii) 0.005 (iii) 0.98

5 (i) $\frac{7}{20}$ (ii) $\frac{11}{20}$ (iii) $\frac{3}{20}$ (iv) $\frac{3}{4}$

6 $\frac{7}{30}$

7 0.8

8 0.6

9 (i) 0.4 (ii) 0.1 (iii) 0.5

10 (i) 0.75 (ii) 0

11 (i) 0.2 (ii) 0.7 (iii) 0.3

12 (i) $\frac{7}{36}$ (ii) $\frac{1}{6}$ (iii) $\frac{5}{18}$ (iv) $\frac{1}{12}$

13 (i) A and D; B and C (ii) 1 (iii) $\frac{1}{3}$

14 (i) 'no heads are obtained'
   (ii) 'at least one head is obtained'; 'fewer than two heads are obtained'

15 (i) No; $(3, 3)$, $(6, 6)$ in both A and B so $P(A \text{ and } B) \neq 0$
   (ii) Yes; if scores are the same then sums are $2, 4, 6, 8, 10, 12$, so sum cannot be 7
   (iii) No; $(1, 6)$, $(6, 1)$, $(3, 4)$, $(4, 3)$ are in both B and C so $P(B \text{ and } C) \neq 0$

Exercise 3d

1 (i) $\frac{1}{3}$ (ii) 0

2 (i) $\frac{9}{38}$ (ii) $\frac{21}{380}$ (iii) $\frac{10}{19}$ (iv) $\frac{39}{95}$

3 (i) $\frac{1}{36}$ (ii) $\frac{5}{18}$ (iii) $\frac{11}{36}$ (iv) $\frac{1}{9}$

4 (i) 0.05 (ii) 0.5

5 (i) $\frac{1}{2704}$ (ii) $\frac{1}{16}$ (iii) $\frac{25}{169}$

6 (i) $\frac{3}{7}$ (ii) $\frac{3}{8}$

7 (i) $\frac{5}{8}$ (ii) $\frac{3}{13}$ (iii) $\frac{1}{20}$

8 (i) No; $P(A) = 0.48 \neq P(A \mid B)$ or $P(B) = 0.3 \neq P(B \mid A)$ or $P(A) \times P(B) \neq P(A \text{ and } B)$
   (ii) 0.66

9 (i) 0.1 (ii) $P(A) \times P(B) \neq P(A \text{ and } B)$
   (iii) $\frac{2}{7}$

10 (i) B and C as total cannot be 7 and 8 at the same time
   (ii) $P(A) \times P(B) = \frac{1}{3} \times \frac{1}{6} = \frac{1}{18} = P(A \text{ and } B)$

11 (i)

	C	C'	Total
Full-time teacher	45	25	70
Part-time teacher	12	18	30
Total	57	43	**100**

   (ii) (a) 0.12 (b) 0.25 (c) 0.82 (d) $\frac{12}{57}$

   (iii) No, $P(C) \times P(F) \neq P(C \text{ and } F)$
   (iv) Full-time teacher, Part-time teacher; Drove a car, Did not drive a car

12 (i) $30 \leqslant \text{age} < 35$ (ii) 24
   (iii) 110 (iv) $\frac{3}{11}$

Exercise 3e

1 (i)

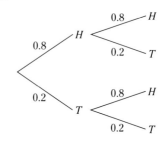

   (ii) 0.64 (iii) 0.96

2 (i)

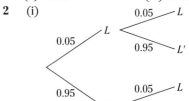

   (ii) (a) 0.0025 (b) 0.095 (c) 0.5

3 (i)

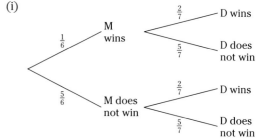

   (ii) (a) $\frac{5}{14}$ (b) $\frac{1}{3}$ (c) $\frac{17}{42}$

4 (i) 0.000625 (ii) 0.04875

5 (i) 0.75 (ii) 0.35 (iii) $\frac{3}{7}$

6 (i) $\frac{5}{12}$ (ii) $\frac{3}{5}$

7 (i) $\frac{15}{38}$ (ii) $\frac{1}{2}$

8 (i) $\frac{5}{8}$ (ii) $\frac{8}{25}$

9 (i)

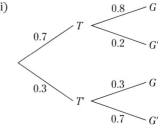

   (ii) 0.65 (iii) $\frac{56}{65}$

10 (i) $\frac{5}{18}$ (ii) $\frac{25}{72}$

11 (a) (i) $\frac{8}{27}$ (ii) $\frac{19}{27}$

    (b) (i) $\frac{5}{21}$ (ii) $\frac{16}{21}$

12 (a) $\frac{1}{4}$

    (b) (i) $\frac{1}{16}$ (ii) $\frac{3}{8}$

    (c) (i) $\frac{27}{64}$ (ii) $\frac{9}{64}$

       (iii) $\frac{5}{32}$ (iv) $\frac{27}{32}$

    (d) $\frac{1}{256}$

13 (i) $\frac{3}{10}$ (ii) $\frac{1}{3}$

14 (a) (i) 0.36   (ii) 0.48

    (b) 0.01024

15 (i) (a) 0.28   (b) 0.54

     (ii) $\frac{47}{110}$

Mixed Exercise 3

1 (i) (a) $\frac{3}{8}$ (b) 0.5

  (ii) $\frac{37}{64}$

2 (i) 0.84   (ii) $\frac{4}{7}$

3 (i)

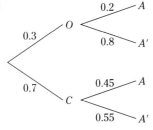

  (ii) 0.375   (iii) 0.16

4 (i) (a) $\frac{1}{816}$ (b) $\frac{5}{68}$ (c) $\frac{55}{272}$

  (ii) $\frac{4}{33}$

5 $\frac{1}{15}$

6 $\frac{135}{181} = 0.746$ (3 s.f.)

7 (i) $\frac{14}{23}$ (ii) 0.226

8 (i) $\frac{7}{9}$ (ii) $\frac{7}{10}$

  (iii) $\frac{2}{9}$

9 (i) 0.364   (ii) 0.086

  (iii) $\frac{18}{43}$

10 (i) $\frac{8}{11}$ (ii) $\frac{3}{11}$

11 (i) $p - 0.3$ (ii) $\dfrac{0.2}{p - 0.3}$

  (iii) 0.7

12 (i) $\frac{3}{44}$ (ii) $\frac{1}{15}$

  (iii) $\frac{3}{11}$

13 (i) $P(E) = \frac{1}{4}$, $P(J) = \frac{1}{2}$, $P(C) = \frac{1}{4}$

(ii)

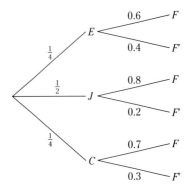

  (iii) $\frac{29}{40}$ (iv) $\frac{3}{11}$

14 (i)

0.68 — C — 0.7 <30
         0.25 30–65
         0.05 >65

0.32 — C' — 0.26 <30
         0.1 30–65
         0.64 >65

  (ii) $\frac{85}{101}$

15 (i)

Time (t minutes)	$2 < t \leqslant 4$	$4 < t \leqslant 6$	$6 < t \leqslant 7$	$7 < t \leqslant 8$	$8 < t \leqslant 10$	$10 < t \leqslant 16$
Frequency	20	44	34	30	30	36

  (ii) 7.55   (iii) $\frac{8040}{18721} = 0.429$ (3 s.f.)

16 (i) 40

  (ii)

[Histogram: Frequency density vs Length of car (m), from 2.80 to 3.40]

  (iii) $\frac{15}{17}$

17   (i) $\frac{1}{24}$ (ii) $\frac{1}{9}$

     (iii) Yes, P(*R* and *Q*) = 0

     (iv) No, P(*R* and *Q*) ≠ P(*R*) × P(*Q*), or
         P(*R* | *Q*) = 0 ≠ P(*R*).

Chapter 4

Exercise 4a

1   (i) 0.1

    (ii)

    (iii) (a) 0.85   (b) 0.55   (c) 0.5   (d) 0.15

2   (i) $\frac{1}{12}$ (ii) $\frac{1}{2}$

3   (i)

r	12	13	14
P(*R* = *r*)	12*k*	13*k*	14*k*

    (ii) $\frac{1}{39}$

4   (i)

x	−1	0	1	3	4	5
P(*X* = *x*)	0.1	0.1	*a*	*a*	0.3	0.1

    (ii) 0.2   (iii) 0.6

5   (i) $\frac{1}{20}$ (ii) $\frac{9}{20}$

6   (i) $\frac{1}{2}$

    (ii)

x	0	1	2
P(*X* = *x*)	$\frac{1}{4}$	$\frac{1}{2}$	$\frac{1}{4}$

7   (ii)

b	0	1	2
P(*B* = *b*)	$\frac{1}{11}$	$\frac{16}{33}$	$\frac{14}{33}$

8   (ii)

x	0	1	2	3
P(*X* = *x*)	0.216	0.432	0.288	0.064

    (iii) 0.352

9   12

10 5

11 (i)

First throw	Second throw					
	1	**1**	**2**	**2**	**2**	**3**
1	2	2	3	3	3	4
1	2	2	3	3	3	4
2	3	3	4	4	4	5
2	3	3	4	4	4	5
2	3	3	4	4	4	5
3	4	4	5	5	5	6

    (ii)

x	2	3	4	5	6
P(*X* = *x*)	$\frac{1}{9}$	$\frac{1}{3}$	$\frac{13}{36}$	$\frac{1}{6}$	$\frac{1}{36}$

12 (i)

First die	Second die			
	1	**2**	**3**	**4**
1	0	1	2	3
2	1	0	1	2
3	2	1	0	1
4	3	2	1	0

    (ii)

d	0	1	2	3
P(*D* = *d*)	$\frac{1}{4}$	$\frac{3}{8}$	$\frac{1}{4}$	$\frac{1}{8}$

    (iii) $\frac{1}{3}$

13 (ii)

x	0	1	2	3	4	5
P(*X* = *x*)	$\frac{1}{6}$	$\frac{5}{18}$	$\frac{2}{9}$	$\frac{1}{6}$	$\frac{1}{9}$	$\frac{1}{18}$

Exercise 4b

1   (i) 0.3

    (ii) 2.9

    (iii) 0.6

2   1.4

3   (i)

x	5	7	8
P(*X* = *x*)	5*k*	7*k*	8*k*

      $k = \frac{1}{20}$

    (ii) 6.9

4   (i) $\frac{1}{2}$

    (ii)

x	0	1	2
P(*X* = *x*)	$\frac{1}{4}$	$\frac{1}{2}$	$\frac{1}{4}$

    (iii) 1

5   (i)

x	0	1	2
P(*X* = *x*)	$\frac{2}{11}$	$\frac{6}{11}$	$\frac{3}{11}$

    (ii) $1\frac{1}{11}$

6   (i) $a + b = 0.5$, $10a + 30b = 7$

    (ii) $a = 0.4$, $b = 0.1$

7   $\frac{1}{3}$

8

x	10	20
P(*X* = *x*)	0.4	0.6

9   (i) $\frac{1}{32}$

    (ii) 1

    (iii) $1\frac{31}{32}$

10 (i)

x	1	2	3
P(*X* = *x*)	$\frac{1}{3}$	$\frac{1}{3}$	$\frac{1}{3}$

    (ii) 2

11 (i)

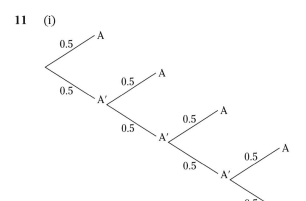

(ii)

x	0	1	2	3	4
P($X = x$)	$\frac{1}{2}$	$\frac{1}{4}$	$\frac{1}{8}$	$\frac{1}{16}$	$\frac{1}{16}$

(iii) $\frac{15}{16}$

Exercise 4c

1 (i) 5.8   (ii) 3.36
2 (i) 0.1   (ii) 0.34, 0.9644
3 (i) 4.2   (ii) $7\frac{1}{3}$ (iii) 3.6696
4 (i) 3, 5, 7, 9, 11, 13; $\frac{1}{9}, \frac{1}{9}, \frac{2}{9}, \frac{2}{9}, \frac{2}{9}, \frac{1}{9}$
     (ii) $8\frac{1}{3}$ (iii) $8\frac{8}{9}$ (iv) 2.98 (3 s.f.)
5 (ii) $1\frac{5}{6}$ (iii) $4\frac{5}{36}$
6 (ii)

x	0	1	2	3
P($X = x$)	$\frac{7}{44}$	$\frac{21}{44}$	$\frac{7}{22}$	$\frac{1}{22}$

     (iii) $1\frac{1}{4}$ (iv) $\frac{105}{176}$
7 (i) 0.2   (ii) 8   (iii) 11.6
8 (i) $a + 2b = 0.78, 2a + 7b = 2.28$; $a = 0.3$,
     $b = 0.24$
     (ii) 5.6016
9 2.56
10 (ii) 7.5   (iii) 185
11 (ii) $\frac{49}{99}$ (iii) $3\frac{58}{99}$ (iv) 1.23 (3 s.f.)

Mixed Exercise 4

1 (ii)

x	0	1	2
P($X = x$)	$\frac{7}{15}$	$\frac{7}{15}$	$\frac{1}{15}$

     (iii) $\frac{3}{5}$
2 (i)

	Die			
	1	2	3	6
Coin H	2	4	6	12
T	1	2	3	6

(iii)

s	1	2	3	4	6	12
P($S = s$)	$\frac{1}{8}$	$\frac{1}{4}$	$\frac{1}{8}$	$\frac{1}{8}$	$\frac{1}{4}$	$\frac{1}{8}$

     (v) 11

3 (i) 0.1   (ii) 1   (iii) 1.75
     (iv) 0.5   (v) 1.2875
4 (i) 1.47, 1.14 (3 s.f.)   (ii) 0.19
5 (i) $\frac{1}{18}$ (ii) $2\frac{7}{9}, 1\frac{14}{81}$ (iii) $\frac{11}{18}$
6 (i) 0.4   (ii) 0.3
     (iii)

l	3	4	5
P($L = l$)	0.1	0.3	0.6

     (iv) 4.5, 0.45
7 (i) $\frac{1}{15}$ (ii) $\frac{8}{15}$ (iii) $\frac{34}{45}$
8 (i) 0.15   (ii) 1.56, 1.4064
9 (i) $\frac{32}{243}$ (ii) 0.0729
     (iii) 0.0100 (3 s.f.)   (iv) $1\frac{2}{3}, 1\frac{1}{9}$
10 (i) $p + q = 0.42, -p + 2q = 0.39$; $p = 0.15$,
     $q = 0.27$
     (ii) 2.5875
11 (i) $\frac{7}{60}$ (ii) $\frac{47}{60}$ (iii) $\frac{40}{47}$
     (iv)

x	0	1	2
P($X = x$)	$\frac{1}{20}$	$\frac{17}{60}$	$\frac{2}{3}$

Chapter 5

Answers are given to 3 s.f. where necessary.

Exercise 5a

1 (i) 0.233   (ii) 0.0368
     (iii) 0.00000590   (iv) 0.0282
2 (i) 0.0231   (ii) 0.208
     (iii) 0.886   (iv) 0.000381
3 (i) 0.102   (ii) 0.143   (iii) 0.000965
4 (i) 0.583   (ii) 0.157
5 (i) 0.452   (ii) 0.414
6 (i) 0.226   (ii) 0.0193   (iii) 0.0193
7 (i) 0.290   (ii) 0.0188
     (iii) 0.159   (iv) 0.745
8 (i) 0.146   (ii) 0.0547
9 (i) 0.0081   (ii) 0.947   (iii) 0.267
10 0.307
11 (i) 0.279   (ii) 0.169
12 (i) $\frac{5}{16}$ (ii) 0.5   (iii) $\frac{3}{16}$

Exercise 5b

1 (i) 0.297   (ii) 0.466
2 (i) 0.455   (ii) 0.545
3 0.914
4 (i) 0.0425   (ii) 0.167
5 (i) 0.323   (ii) $\frac{1}{36}$
6 0.766
7 (i) 0.6   (ii) 0.346
8 (i) $\frac{2}{3}$ (ii) 12

9 (i) 0.925   (ii) 17
10 (ii) 0.329

Exercise 5c

1 (i) 5.04   (ii) 3.2256   (iii) 1.80
2 0.180
3 (i) 3
    (ii) $P(X = 2) = 0.229$, $P(X = 3) = 0.243$ (highest probability), $P(X = 4) = 0.182$
4 (i) 23   (ii) 1.92
5 (i) 8   (ii) 0.0467   (iii) 1.30
6 (i) 5, 1.58   (ii) 0.01074...
    (iii) 0.01074...   (iv) 0.0215
7 (i) $\frac{1}{4}$ (ii) 5, 1.94   (iii) 0.561
8 (i) 0.0081   (ii) 1.6   (iii) 0.84
9 (i) 0.3   (ii) 8   (iii) 0.0100
10 (i) 9, 0.4   (ii) 0.232
11 (i) 0.1   (ii) 0.354
    (iii) 6   (iv) 25
12 (i) 0.216, 0.288   (ii) 1.2, 0.72   (iii) 1.2, 0.72

Mixed Exercise 5

1 (i) 0.183   (ii) 0.00204   (iii) 0.200
2 (i) 0.0173   (ii) 0.118
3 (i) 0.0638   (ii) 0.465   (iii) 0.267
4 (i) 0.247   (ii) 0.144
5 (i) 4.05   (ii) 1.49
6 (i) 2.4   (ii) 0.439   (iii) 2.04
7 (i) 0.0134   (ii) 14   (iii) 8
8 (i) 0.4   (ii) 4
9 (i) 0.624   (ii) 2.5   (iii) 1.275
10 (i) 0.0860   (ii) 6
11 (i) 0.0749   (ii) 6.75
12 0.737

Chapter 6

Exercise 6a

1 (i) 0.8089   (ii) 0.8089
    (iii) 0.1911   (iv) 0.1911
2 (i) 0.0359   (ii) 0.2578
    (iii) 0.9931   (iv) 0.9131
    (v) 0.0049   (vi) 0.9911
    (vii) 0.9686   (viii) 0.2343
    (ix) 0.0312   (x) 0.9484
    (xi) 0.9803   (xii) 0.0021
3 (i) 0.05   (ii) 0.05
    (iii) $0.0999 \approx 0.1$ (iv) 0.025
    (v) 0.995   (vi) 0.01
    (vii) 0.0025   (viii) 0.975

4 (i) 0.044   (ii) 0.8185
    (iii) 0.3023   (iv) 0.1336
5 (i) 0.1703   (ii) 0.5481
    (iii) 0.3639   (iv) 0.4582
    (v) 0.4798   (vi) 0.9624
    (vii) 0.0337   (viii) 0.9082
    (ix) 0.2729   (x) 0.0306
6 (i) 0.925   (ii) 0.9
7 $49.98\% \approx 50\%$
8 (i) 0.9   (ii) 0.7
9 (i) 0.55   (ii) 0.15
10 (i) 0.9   (ii) 0.1

Exercise 6b

1 (i) 0.0668   (ii) 0.4013   (iii) 0.1747
2 (i) 0.7190   (ii) 0.0618   (iii) 0.4621
    (iv) 0.0456
3 (i) 0.0548   (ii) 0.1448   (iii) 0.9544
4 (i) 0.0106   (ii) 0.9857
5 (i) 0.3015   (ii) 0.5231   (iii) 0.3792
6 (a) (i) 0.8413   (ii) 0.9257
    (b) 207
7 0.00003844
8 (i) 0.6554   (ii) 0.1171...   (iii) $8.2 \approx 8$
9 (i) 0.1056   (ii) 0.7734   (iii) 0.6678
10 Size 1: 0.159, Size 2: 0.775, Size 3: 0.067
11 0.8664
12 (i) 2   (ii) 454   (iii) 124
    (iv) 47

Exercise 6c

1 (i) 0.015   (ii) 0.796   (iii) -1.887
    (iv) -0.454 (v) -0.562 (vi) 1.019
2 (i) 1.940   (ii) -0.695 (iii) -0.915
    (iv) 0.722   (v) 1.036

Exercise 6d

1 (i) 73.85   (ii) 4.65
    (iii) 190.742   (iv) 1.468
2 (i) 176.578   (ii) 166.723
3 (i) 625.6 g   (ii) 567.1 g
4 (i) (384.32, 415.68)   (ii) 10.784
5 (i) 0.383   (ii) 96.74
6 (125.5 h, 194.5 h)
7 (i) 0.4839   (ii) 96.92, 103.08

Exercise 6e

1 40
2 10.73
3 (i) 8.31   (ii) 35.9%
4 (i) 35.5   (ii) 0.0907
5 1.75 m
6 2.74 kg, 2.78 kg
7 52.73, 11.96

8 (i) 6.99, 0.105   (ii) 0.0105
9 39.5, 5.32
10 (i) 0.321   (ii) 14.25
11 0.203
12 (i) 92.7%   (ii) 1.32   (iii) 1.7%
13 5.102 g
14 (i) 58.79, 3.304   (ii) 4.46
15 3040 km
16 (i) 828.45 g   (ii) 0.615 g
17 (i) 0.507   (ii) 0.00258
18 535.1 g, 12.22 g
19 (i) 109 km/h   (ii) 0.428
20 (i) 0.9332   (ii) $\mu = 0.9\sigma$

Exercise 6f

1 P(2.5 < X < 9.5)   **2** P(3.5 < X < 8.5)
3 P(10.5 < X < 24.5)   **4** P(1.5 < X < 7.5)
5 P(X > 24.5)   **6** P(X > 75.5)
7 P(X < 108.5)   **8** P(X < 45.5)
9 P(66.5 < X < 67.5)   **10** P(19.5 < X < 20.5)

Exercise 6g

1 0.1958
2 (i) 9.5, 5.89   (ii) $np = 9.5 > 5, nq = 15.5 > 5$
    (iii) (a) 0.4932   (b) 0.0968
        (c) 0.1082   (d) 0.6598
3 (i) 0.0154   (ii) 0.8145   (iii) 0.02
4 0.1174
5 (i) 0.678
    (ii) (a) 0.0318   (b) 0.8345
6 (i) 0.9474   (ii) 0.6325   (iii) 0.5914
    (iv) 0.0111
7 (i) 0.4502   (ii) 0.0996   (iii) 0.484
8 (i) (a) 0.2304   (b) 0.9222
    (ii) 0.8531
9 0.1432
10 (i) 0.1853   (ii) 0.1838   (iii) 0.81%
11 (i) 0.0829   (ii) 0.2749
12 (i) 0.0014
    (ii) (a) 0.0255   (b) 0.5742
13 (i) 0.117   (ii) 0.00361   (iii) 0.556
14 0.8871
15 0.8724

Mixed Exercise 6

1 (i) 72.00   (ii) 0.370
2 (i) 0.0088   (ii) 7.84; 42.16, 57.84
3 (i) 0.0276   (ii) 7.72
4 (i) 0.0036   (ii) 0.110
5 (i) 16.2, 1.67   (ii) 0.274
6 (i) 0.2025   (ii) 480.8 g
7 (i) 8.752   (ii) 0.5458
8 (i) 7.29   (ii) 0.136   (iii) 0.370
9 (i) 0.298   (ii) 0.118   (iii) 13

10 (i) 2.23   (ii) 0.6542   (iii) 0.112
    (iv) 0.2495
11 (i) 0.3735   (ii) 0.0419
    (iii) Does not agree with model;
        not symmetrical, so not normal
12 (i) 0.398   (ii) 9   (iii) 0.9717
13 (i) 108.42, 13.4   (ii) 0.4309

Sample Paper 1

1 (i) $\frac{6}{25}$ (ii) $\frac{6}{13}$
2 (i) 0.325   (ii) 0.842
3 (i)

x	0	1	2
$P(X = x)$	0.1	0.6	0.3

(ii) 1.2, 0.36
4 (i) 0.408   (ii) 144   (iii) 1440
5 (i) (a) 0.0548   (b) 0.0297
    (ii) 0.970
6 (i) (a) 27.65, 3.84   (b) Data given in classes
    (ii)

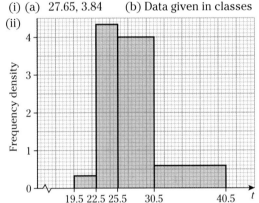

(iii) 20th & 21st values in 26–30 class.

Sample Paper 2

1 19.51, 0.254
2 90.6, 23.1
3 (i) $\frac{1}{10}, \frac{1}{5}, \frac{3}{10}$ (ii) 4
4 (i) 43   (ii) $p = 3$ or $4, q = 4$
    (iii) E.g.:

```
2 | 3
3 | 2 6
4 | 1 3 5
5 | 3 3 4 6 7
6 | 0 1 ① 2 3
7 | 1 5
```

(iv) E.g.: Shows all the data
    (v) E.g.: Shows the IQR
5 (i) 0.0608   (ii) 0.220
6 (i) (a) 48   (b) 432
    (ii) 55
7 (ii) 0.856   (iii) 0.332   (iv) 0.747

Index

TV SPORT STARS OF
FORMULA ONE
ANNUAL 2013

Written by Jon Culley
Design by Nicky Regan

£7.99

CONTENTS

The success of Lewis Hamilton and Jenson Button has confirmed Great Britain's status as the biggest producer of world champion Formula One drivers. The championship has now been won by 10 British drivers, sharing 14 titles. These are their stories.

GREAT BRITONS

THE STORY OF BRITAIN'S 10 FORMULA ONE WORLD CHAMPIONS

MIKE HAWTHORN
1958

In 1958 the Formula One drivers' championship went to a British driver for the first time, breaking the dominance of Italy's Giuseppe Farina and Alberto Ascari, and the brilliant Argentine, Juan Manuel Fangio, who had won it five times but was effectively retired.

The flamboyant, public school-educated Hawthorn, whose father ran a garage at Farnham, Surrey, had won the French Grand Prix in 1953 and the Le Mans 24 Hours in 1955. He joined Ferrari in 1957 and was crowned world champion in 1958, despite winning only one race compared with four by fellow Brit Stirling Moss.

In the final race, in Morocco, Moss would be champion if he could win and clock the fastest lap, provided Hawthorn finished no better than third.

Moss duly did win and set the fastest lap, but the attempt by his Vanwall team-mates, Tony Brooks and Stuart Lewis-Evans, to keep Hawthorn out of the first three failed when both broke down, allowing Hawthorn to finish second and clinch the title by a single point.

Sadly, Lewis-Evans died after his car caught fire, capping a tragic year in which Hawthorn's Ferrari team-mates, Peter Collins and Luigi Musso, were killed in separate crashes. Hawthorn himself died the following year, not on the track but in a road accident near his home.

GRAHAM HILL
1962 & 1968

By the early sixties, only Ferrari challenged the British-built cars in Formula One and a golden age for British drivers was about to begin.

With the retirement of Tony Brooks and Stirling Moss effectively to quit after a serious accident, the way was clear for two more Brits, Graham Hill and Jim Clark, to turn the 1962 championship into a two-horse race.

Hill, a personality known for his ready wit and distinctive moustache, began the 1962 season in a much-improved BRM against Clark's revolutionary monocoque Lotus 25. After three wins each, the final event in South Africa became a title decider. Clark took pole and clocked the quickest lap only to be forced off by an oil leak, leaving Hill to clinch the crown.

It would not be Hill's last title. He and Clark, by now a double world champion, became team-mates when Hill rejoined Lotus and at the 1968 South African Clark beat Hill in a Lotus 1-2. But tragically Clark was killed in a crash at Hockenheim and never raced in Formula One again. Hill won three of the remaining 11 GPs and was crowned world champion in Mexico in November. Hill – the only driver to win the 24 Hours of Le Mans, the Indianapolis 500 and the Formula One world championship – died when his light aircraft crashed in fog in Hertfordshire in 1975.

JIM CLARK
1963 & 1965

Born into a Scottish farming family, Clark raced first in rallies and hill climbs near his home in Berwickshire, as a hobby. But his talent shone after he joined a local club and he won a string of national events.

In 1958 he finished second in a GT race at Brands Hatch won by Lotus founder Colin Chapman, who asked 22-year-old Clark to drive in the Formula Junior series.

Within two years, Clark had graduated to Formula One and quickly rose to the top. Second in the 1962 world championship, he broke records in 1963, winning five of the first seven races in his Lotus 25 to be confirmed champion with an unprecedented three races to spare.

He landed his second world championship in 1965, winning six of the first seven races and missing the Monaco GP only because he was away winning the Indianapolis 500, the first non-American victor in almost half a century.

When he won the South African GP on New Year's Day, 1968, he overtook Juan Miguel Fangio's 24 GP wins to become the most successful racer of all time to that date. But fate was to make it his last Formula One race. Four months later, in a Formula Two race in Hockenheim, he died when his car left the track and hit a tree.

JOHN SURTEES
1964

Any one of three British drivers could have won the 1964 championship as Graham Hill (39pts) arrived for the season finale in Mexico, leading from John Surtees (34) with title-holder Jim Clark on 32 in spite of winning three races to his rivals' two each.

To retain his crown, Clark needed to win and hope that 1962 champion Hill would finish no higher than third. A top-three finish would give Hill the title regardless of where Surtees finished, so long as Clark did not win, in which case Surtees could win by finishing second or higher, provided Hill was not in the top three.

Ferrari driver Surtees was bidding to be the first man to be world champion on two wheels and four, having won seven world motorcycle titles, but seemed to have no chance, trailing in fifth place as Clark led from the American, Dan Gurney, while Hill battled with another Ferrari driver, Lorenzo Bandini, for third.

But a collision with Bandini left Hill's car damaged and then Clark's engine seized, leaving Gurney in front with Bandini second. Hill had dropped back but would be champion so long as Surtees did not finish in the top two. Ferrari knew this and signalled Bandini to move over, allowing Surtees to cross the line with only Gurney ahead of him and take the title by a point.

JACKIE STEWART
1969, 1971 & 1973

Stewart, the son of a Scottish garage proprietor who had excelled at clay pigeon shooting as well as at the wheel of a car, became the fifth British driver to win the drivers' championship when the French-built Matra MS80 developed by Ken Tyrrell came home first in a blanket finish to the Italian GP at Monza, giving him the title after eight of the 11 races.

A brilliant and courageous racer, Stewart would go on to be motorsport's first global superstar, taking on the appearance of a rock star with his long hair. His image helped popularise motor racing and made him wealthy, yet he was never a playboy, always a devoted family man and an eloquent, determined campaigner for safety on the track, having escaped unhurt from a serious crash himself in 1966 but seen close friends Piers Courage and Jochen Rindt die in 1970.

He won easily again in 1971, winning five of the first seven races and clinching the title again in race eight of 11 despite failing to finish because of mechanical problems.

In 1973, seeing off the challenge of the Team Lotus pair, Emerson Fittipaldi and Ronnie Peterson, he took title number three of his six-year association with Tyrrell but brought forward his retirement when team-mate Francois Cevert died in practice for the United States GP.

JAMES HUNT
1976

Where Jackie Stewart was never the playboy, James Hunt was the genuine article. Introduced to Formula One with the support of the outrageous Lord Alexander Hesketh, the well-spoken, glamorous Hunt partied long and hard and had a succession of beautiful girlfriends. He would die from a heart attack at the age of only 45.

His nickname 'Hunt the Shunt' said everything about his driving style but he had talent, as demonstrated in Holland in 1975 when he beat Niki Lauda, with whom he would fight a thrilling duel in 1976.

Hesketh ran out of money and Hunt was out of a job until an unexpected and fortuitous vacancy came up at McLaren. Hunt won in Spain and France and delighted the home crowd by winning at Brands Hatch, although he would later be disqualified. Lauda, nonetheless, had a big lead – but then came a crash at the Nürburgring that almost claimed the Austrian's life.

Hunt won the German race and again in Holland but despite disfiguring burns Lauda made a miraculous comeback and, awarded Hunt's British GP points, led by 64 points to 47 with three races left. However, two more victories put Hunt three points in front going into the last round in Japan and when Lauda dropped out with the track awash with heavy rain, Hunt stayed on to finish third and take the title.

NIGEL MANSELL
1992

When Nigel Mansell became Britain's seventh world champion in 1992 – the first for 16 years – it was the joyous reward for 13 years of perseverance. An aggressive, exciting driver, Mansell crashed more often than he won but still won 31 times, making him the most successful British driver of all time.

Mansell had to pay his own way through the ranks and twice suffered injuries that might have left him paralysed. He fought back to land a Formula One drive with Lotus but never finished higher than third in four seasons. His fortunes rose when he left Lotus for Williams, where he finished runner-up for the title twice in a row, narrowly pipped by McLaren's Alain Prost in 1986, then losing out to team-mate Nelson Piquet in 1987.

After a poor 1988, Mansell accepted an offer to join Ferrari, for whom he won on his debut in Brazil and finished fourth. The following season, disillusioned after Prost joined as lead driver, Mansell threatened to retire, but deferred doing so after Williams took him on again. At last Mansell realised his dream. Back to winning ways in the FW14 car, he finished second in 1991 and took the title in 1992 with nine victories, including the first five of the season, putting him so far ahead he was confirmed champion after only 11 of the 16 races.

DAMON HILL
1996

Nigel Mansell left to race IndyCars in America after his 1992 title triumph, after which Williams replaced him with Damon Hill, the son of the late Graham Hill, twice champion in Britain's golden age of the 1960s.

Damon had never won races consistently in any formula but impressed Williams as a test driver, and helped develop Mansell's title-winning car. Hill's name had not given him an easy passage. His father's death, in fact, left the family in reduced circumstances.

Damon finished third to team-mate Alain Prost in his first season in Formula One, winning three times. Prost retired, and when his replacement, Ayrton Senna, was killed in only his third race for Williams, Hill suddenly found himself leader of a team in mourning. Yet he took on the responsibility bravely and finished second only to Michael Schumacher two seasons in a row, on the first occasion amid controversy after the German's car collided with Hill's in the final race.

Even so, there was talk that Hill should be doing better with the car acknowledged as fastest on the grid. His performance in 1996 ought to have put paid to all doubts. Never off the front row, Hill won eight races and the title, becoming the first son of a former champion to triumph. Yet Williams had already decided to replace him in 1997.

LEWIS HAMILTON
2008

The ninth British world champion, Lewis Hamilton could not have left it later to clinch the extra point he needed to finish in front of Felipe Massa, overtaking Timo Glock on the final bend of the final lap of the final race to come home fifth, taking four points instead of three to finish on 98 points to Massa's 97.

Hamilton was 10 points behind Massa after eight races, but back-to-back wins at Silverstone and Hockenheim put him four in front. It was a lead he did not surrender, but with Massa winning two more races and Hamilton without another victory until the penultimate round in China, race 18 arrived with the gap just seven points and Massa with the advantage of racing on his home track, Interlagos.

If Hamilton finished sixth or worse, a win for Massa would clinch the title. Massa mastered wet conditions perfectly, starting on pole and leading throughout. Hamilton, fourth on the grid, was fourth in the race until a tyre change on lap 66 of the 71 allowed Glock to pass him. Then, in heavier rain, a Hamilton error let Sebastian Vettel through. As Massa crossed the finish line, Ferrari began to celebrate but in the 38.907 seconds that passed before Hamilton crossed the line, the British driver overtook a faltering Glock to regain fifth place on the final bend.

JENSON BUTTON
2009

Button's victory in the 2009 championship was almost as much a fairytale as Hamilton's the year before. Not that Button was a newcomer. He had been racing in Formula One for nine years and while he had a superb season to finish third overall in 2004, he had won only one race, the 2006 Hungarian GP, in more than 150 attempts.

But that was to change dramatically after Honda's withdrawal from Formula One in 2008 led to a buy-out by team principal Ross Brawn, which kept Button employed. Some innovative technical changes gave the Mercedes-powered Brawn GP cars an advantage he exploited to spectacular effect, winning six of the first seven races.

He did not win again, driving more cautiously as his lead came under pressure. But he failed to score points only once, steadily accumulating enough to stay at the top of the standings all season, even as the likes of Sebastian Vettel and Rubens Barrichello made up ground.

He clinched the title in Brazil in the last race but one. Only 14th on the grid, he plotted a path through a series of incidents on the first lap to move up to finish fifth, which was enough to take the title and secure the Constructor's Championship for Brawn.

GREAT BRITONS

BORN: July 3, 1987, Heppenheim, Germany

2012 TEAM: Red Bull

DEBUT: 2007 United States GP (BMW Sauber)

FIRST FORMULA ONE WIN: 2008 Italian GP (Toro Rosso)

FIRST FORMULA ONE TITLE: 2010 (Red Bull)

SEBASTIAN VETTEL

DRIVER PROFILES 1

SEBASTIAN VETTEL FIRST BEGAN TO ATTRACT SERIOUS ATTENTION IN THE GERMAN BMW ADAC SERIES, FINISHING 2ND IN 2003 AND DOMINATING IN 2004, WHEN HE WAS CHAMPION WITH A STAGGERING 18 WINS FROM 20 RACES.

In 2005 he stepped up to F3 Euro Series, finishing 5th behind Lewis Hamilton and runner-up to another Brit, Paul di Resta, in 2006. He began 2007 in World Series by Renault and was given his Formula One break at the US GP in Indianapolis the same year when replacing the injured BMW Sauber driver Robert Kubica and finishing 8th, making him Formula One's youngest ever points scorer at 19 years old.

Switching to Toro Rosso, he achieved his and their first win in September, 2008, at Monza, at 21 years and 73 days the youngest winning driver in GP history. That victory earned him the Red Bull race seat for 2009, when he narrowly lost out to Jenson Button in a thrilling title race. The progression to world champion was almost inevitable, although it was not until the final race, in Abu Dhabi, of another exciting summer that he clinched it, overtaking Fernando Alonso and team-mate Mark Webber in the standings after winning the race from pole, establishing another record as youngest-ever drivers' champion at 23.

In 2011 he simply dominated the Formula One series, claiming pole a record 15 times and winning 11 races in the RB7, giving him the title with four rounds to spare.

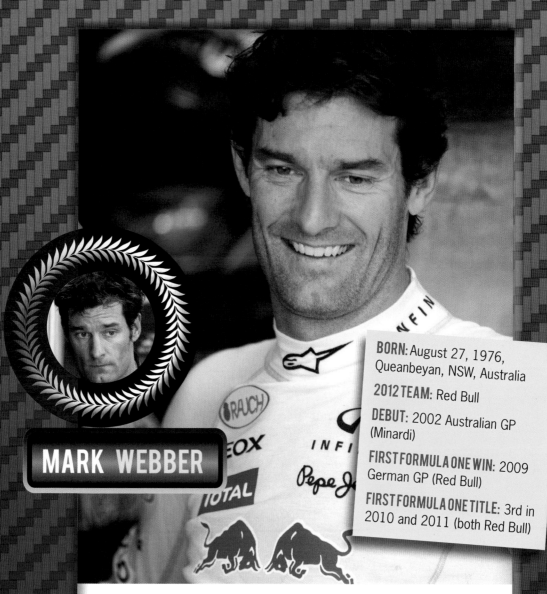

BORN: August 27, 1976, Queanbeyan, NSW, Australia

2012 TEAM: Red Bull

DEBUT: 2002 Australian GP (Minardi)

FIRST FORMULA ONE WIN: 2009 German GP (Red Bull)

FIRST FORMULA ONE TITLE: 3rd in 2010 and 2011 (both Red Bull)

MARK WEBBER

MARK WEBBER'S CAREER BEGAN BACK IN 1994, WHEN HE ENTERED THE AUSTRALIAN FORMULA FORD CHAMPIONSHIP AND FINISHED 14TH. HE MOVED TO BRITAIN THE FOLLOWING YEAR AND CAUGHT THEIR EYE BY WINNING THE FORMULA FORD FESTIVAL FOR VAN DIEMEN IN 1996.

He graduated to F3 and acquitted himself well but a lack of cash delayed his promotion to Formula 3000 and he spent two seasons in GT racing. When the chance came, though, Webber grabbed it with both hands and two impressive seasons in Formula 3000 – he was 3rd in 2000 and 2nd in 2001 – landed him his Formula One debut with Minardi. Success initially proved elusive: two seasons with Jaguar Racing cemented his position as a potential star but he then turned down Renault – champions in 2005 and 2006 – in favour of Williams, who struggled to compete.

He signed for Red Bull in 2007, achieving the team's first-ever front row grid slot at Silverstone in 2008 and landing his first Formula One win at the Nürburgring in 2009. He won in Brazil the same year and in 2010 had a chance to become the first Australian champion since Alan Jones in 1980. He led the standings at half-way following wins in Spain, Monaco, Britain and Hungary but lost out in the four-way showdown at the end of the season and finished 3rd overall, a position he occupied again as team-mate Sebastian Vettel took charge in 2011.

DRIVER PROFILES 2

BORN: July 29, 1981, Oviedo, Spain

2012 TEAM: Ferrari

DEBUT: 2001 Australian GP (Minardi)

FIRST FORMULA ONE WIN: 2003 Hungarian GP (Renault)

FIRST FORMULA ONE TITLE: 2005 (Renault)

FERNANDO ALONSO

WORLD KARTING CHAMPION AT ONLY 15, ALONSO MADE HIS FORMULA ONE DEBUT FOR MINARDI AS A 19-YEAR-OLD AT THE 2001 AUSTRALIAN GP. SWITCHING TO RENAULT IN 2003, HE BEGAN A METEORIC RISE, BECOMING THE YOUNGEST RACE WINNER IN HUNGARY THAT YEAR AND, IN 2005, THE YOUNGEST WORLD CHAMPION AT 24.

He retained his title in 2006, winning six of the first nine races before holding off a challenge from Michael Schumacher. Switching to McLaren, he was only just pipped to the 2007 title, finishing a point behind champion Kimi Räikkönen and level with team-mate Lewis Hamilton. But he had a turbulent relationship with Hamilton and the year was marred by a spying scandal involving McLaren and Ferrari.

Alonso, released from his contract, returned to Renault. He exceeded expectations by finishing fifth in 2008 but 2009 was less successful and in 2010 he replaced Räikkönen at Ferrari. Having begun in style by winning the opening race in Bahrain GP, he had to wait until Germany for a second victory but then took the Italian, Singapore and Korean races to be in contention for the championship going into the final race in Abu Dhabi, where he led the standings only for strategic errors to allow Sebastian Vettel to snatch the title. After that near-miss, 2011 was disappointing for Alonso and the team, with a solitary victory in the British GP.

BORN: April 25, 1981, São Paulo, Brazil

2012 TEAM: Ferrari

DEBUT: 2002 Australian GP (Sauber)

FIRST FORMULA ONE WIN: 2006 Turkish GP (Ferrari)

BEST FORMULA ONE SEASON: 2nd in 2008 (Ferrari)

FELIPE MASSA

MASSA FIRST RACED FOR FERRARI IN 2006 BUT HIS LINKS WITH THE ITALIAN MARQUE GO BACK TO 2003 THE YEAR AFTER HIS FORMULA ONE DEBUT WHEN HE SPENT A YEAR AS FERRARI TEST DRIVER ON THE ADVICE OF PETER SAUBER, WHO THOUGHT HE NEEDED TO ADD SOME MATURITY TO HIS UNDOUBTED TALENT.

Back racing in 2004, he had two respectable years with Sauber before joining Ferrari in 2006. He performed well, pushing team-mate Michael Schumacher, claiming his maiden victory in Turkey and becoming the first Brazilian since Ayrton Senna to win his home GP in São Paulo. He finished 3rd in the 2006 championship.

Frustratingly for him, the 2007 title went to his new Ferrari team-mate, the Finn Kimi Räikkönen, while he finished 4th – yet he emerged as Ferrari's No1 driver in 2008. Indeed, after five wins he was in contention for the title himself going into a dramatic season finale in Brazil that brought high drama. Massa won the race for the second time and believed he was champion as he crossed the line, only for Lewis Hamilton to snatch fifth place only seconds from the finish to take the title by a point.

It would prove Massa's last race win before suffering life-threatening head injuries in practice for the Hungarian GP the following year. Happily he recovered but played second fiddle to team-mate Fernando Alonso in 2010 and 2011, each time finishing sixth.

01 GERMANY

VENUE: Nürburgring GP-Strecke, Nurburg

CIRCUIT LENGTHS: 5.148km

TURNS: 16

LAPS: 60

THE ICONIC NÜRBURGRING IN GERMANY HOSTS THE NATION'S GRAND PRIX EVERY OTHER YEAR, AS WELL AS BEING A TESTING FACILITY FOR VIRTUALLY EVERY PRODUCTION CAR.

Only a section of the track is now used for Formula One racing whereas in the early years of the German Grand Prix the entire 25.9km circuit was used for races. From 1931-1966 the Nordschleife section was the Formula One circuit and is now usually the track used to ascertain the racing pace of new cars.

The first Grand Prix in the Nürburgring's current form, which has been widely criticised for lacking the character of the wider circuit, was in 1985 and the track has also been the venue for the European Grand Prix on 11 occasions. Germany's Michael Schumacher has the most Formula One wins at the circuit, with five, while British drivers have an excellent record at the circuit.

Jackie Stewart and John Surtees won there five times between them while the last Formula One race at the Nürburgring was won by Lewis Hamilton in 2011. The popular British driver, Johnny Herbert, also secured the third and final race win of his career in 1999 at the Nürburgring, winning the European Grand Prix for the Stewart-Ford team.

DID YOU KNOW?

The 'S' turn on the new GP-Strecke section was renamed after Michael Schumacher in 2006 following his retirement. In 2011, a year after he returned to Formula One, he became the first man in Formula One to drive through a corner named after him.

02 BRAZIL

VENUE: Interlagos, Sao Paulo

CIRCUIT LENGTHS: 4.309km

TURNS: 15

LAPS: 71

OOZING CHARACTER, INTERLAGOS IN SÃO PAULO HAS RETAINED ITS STATUS AS ONE OF FORMULA ONE'S MOST EXCITING TRACKS DESPITE HAVING BEEN RADICALLY ALTERED IN THE 1980S.

When Formula One racing first came to Brazil, the Interlagos circuit was eight kilometres long, almost double its current length. Due to safety concerns including bumpy sections, inadequate barriers and deep ditches, Interlagos lost the Brazilian Grand Prix for ten years, reclaiming it only in 1990 after extensive redevelopments that included the shortening of the circuit.

The Brazilian race has often been the last in the calendar, with many Formula One champions being crowned at Interlagos. Lewis Hamilton's maiden title in 2008 was clinched in thrilling fashion in Brazil, Felipe Massa winning the race but being denied the championship after he had finished when Hamilton overtook Timo Glock.

There have been nine Brazilian winners of their own Grand Prix, although Nelson Piquet's brace of victories were in Rio rather than at Interlagos. Michael Schumacher has won the most races at Interlagos with four. Alain Prost won the Brazilian Grand Prix six times but five of the Frenchman's successes were at Rio's Jacarepaguá circuit.

DID YOU KNOW?

While universally known by its traditional title Interlagos, the venue is officially called the 'Autódromo José Carlos Pace', after the Brazilian racing driver Carlos Pace – a former Interlagos winner whose career was tragically cut short when he was killed in a plane crash in 1977.

03 AUSTRALIA

VENUE: Albert Park, Melbourne

CIRCUIT LENGTHS: 5.303km

TURNS: 16

LAPS: 58

USUALLY THE CURTAIN-RAISER OF THE FORMULA ONE SEASON, THE AUSTRALIAN GRAND PRIX, HELD IN MELBOURNE'S ALBERT PARK SINCE 1996, HAS BECOME A PRIME HUNTING GROUND FOR BRITS.

Jenson Button has won three times down under, the latest coming last year, David Coulthard has been first twice at Melbourne, with Lewis Hamilton and Damon Hill each recording victories as well.

Michael Schumacher boasts the most wins at Albert Park, triumphing four times during his Ferrari days, and the German also still has the record for the fastest racing lap time in Melbourne, with a 1'24.125 in 2004, although his compatriot Sebastian Vettel recorded a quicker qualifying time in 2011. The track circles Albert Park Lake, with sections of it usually open to public vehicles although, because the entire track was rebuilt in preparation for Formula One racing, the surfaces of these public stretches are still smooth.

Considered an easy track to navigate, because of its consistent and regular corners, overtaking is not easy because there are few lengthy straights.

DID YOU KNOW?

The Australian Grand Prix was held in Adelaide when the race was incorporated into the Formula One calendar in 1985 and, rather than being the first race of the year, it was the last for each of the 11 years the event was held in South Australia.

04 MALAYSIA

VENUE: Sepang International Circuit, Sepang

CIRCUIT LENGTHS: 5.543km

TURNS: 15

LAPS: 56

A VERY COMPLICATED BUT ALSO VERY POPULAR TRACK AMONGST THE DRIVERS, SEPANG IN MALAYSIA HAS BEEN A FEATURE OF FORMULA ONE SINCE 1999.

With long high-speed straights and tight twisting complexes, Sepang is a challenging circuit but one that is conducive to exciting racing, with the very wide track perfect for overtaking.

FIA-sanctioned motor racing in Malaysia dates back to 1962 but it was the emphasis in the 1990s on using the auto trade to help Malaysia industrialise that contributed to the fabulous facility at Sepang being built and Formula One coming to the country.

Britain's Eddie Irvine won the first Malaysian Grand Prix for Ferrari and the Italian giants have found Sepang a happy hunting ground, winning six of the 13 races held there. Three of those were won by Michael Schumacher the equal highest number by an individual at Sepang. Current Ferrari driver Fernando Alonso also has three Malaysia wins to his name although only last year's came in the red of Ferrari.

DID YOU KNOW?

The 2001 Malaysian race was affected by extremely wet weather and the Ferraris of Michael Schumacher and Rubens Barrichello spun off almost simultaneously on lap three but remarkably the duo recovered to record a "1-2" for Ferrari.

HE DOES NOT HAVE AN ENTRY IN WHO'S WHO BUT ANY DIRECTORY OF THE MOST INFLUENTIAL FIGURES IN THE WORLD OF SPORT WOULD BE INCOMPLETE WITHOUT A SECTION ON BERNIE ECCLESTONE.

Instantly recognisable for his flop of white hair and tinted glasses, the octogenarian multi-billionaire has been almost single-handedly responsible for making Formula One the enormous commercial success it has become. Today, as president and chief executive of Formula One Management and Formula One Administration, and through his part-ownership of Alpha Prema, the parent company of the Formula One Group of companies, he effectively owns the sport.

Ecclestone was practically born an entrepreneur. The son of a trawlerman from Bungay in Suffolk, he made money in the school playground by selling cakes bought with the pay from his paper rounds, moving on quickly to pens and watches. After his family moved to the Kent fringes of London, Ecclestone left school at 16 and took a job at a local gasworks, but continued to trade, this time in spare parts for motorcycles, in which he had a keen interest. He raced himself for a while, but business was always his strong suit. In partnership with a friend, Fred Compton, he started Compton & Ecclestone, which became one of Britain's biggest motorcycle dealerships.

His empire had expanded to include property and a car auction company by the time he made his first direct investment in Formula One in 1957, buying the struggling Connaught team, for whom he drove himself at Monaco, although he failed to qualify. He went into management, too, although he was so shaken when his driver, the Welshman Stuart Lewis-Evans, died from burns sustained in a crash in Morocco that he quit the sport. His second involvement, as manager of a Lotus Formula Two team featuring Jochen Rindt and Graham Hill, was also darkened by tragedy when Rindt was killed at Monza.

Yet Ecclestone came back again in 1972 when he bought Brabham, restoring the name's prestige in the early 80s when Nelson Piquet won the drivers' championship twice.

Ecclestone's success enabled him to influence the strategic direction of the sport. With the help of lawyer and motor racing enthusiast Max Mosley, who owned the March team, he turned the Formula One Constructors' Association (FOCA) into the most powerful grouping in the sport. Spotting the potential of television coverage for generating revenue, he won the right for FOCA to negotiate with broadcasters on behalf of the teams. He secured lucrative global TV deals handing just under half the TV income back to FOCA, just less than a third to the sport's governing body, the FIA, and retaining the rest for his own Formula One Promotions and Administration, who provided the prize money.

Ecclestone's shrewd dealings generated enormous prosperity for the sport and a personal wealth that put him in the domain of the super-rich, although he did not flaunt his fortune and would have happily stayed out of the public eye. However, the furore over his £1million donation to Tony Blair's New Labour party, which became linked to motor racing's special treatment over the ban on tobacco advertising in sport, thrust him into the headlines. Lord Neill, the parliamentary standards watchdog, ultimately advised Labour to return the money.

He found himself the subject of unwanted attention again in 2010 when he and his girlfriend, Fabiana Flosi, were attacked on a London street. A group of men stole jewellery from them said to be worth £200,000 and Ecclestone, who was knocked out, was left with a large black eye.

Married twice and with a personal fortune estimated at £2.5billion in the 2012 Sunday Times Rich List, Ecclestone underwent a triple coronary bypass in 1999 but shows no signs of relinquishing his control of the sport.

MR FORMULA ONE – A PROFILE OF BERNIE ECCLESTONE

MICHAEL SCHUMACHER

BRILLIANT YET OFTEN CONTROVERSIAL FOR HIS DRIVING TACTICS, MICHAEL SCHUMACHER IS WITHOUT QUESTION ONE OF THE FINEST DRIVERS TO GRACE FORMULA ONE.

The winner of 91 Grand Prix races and seven world championships more than any other driver in Formula One history Schumacher retired at the end of 2006. His return in 2010 may have been a mistake. Yet nothing can take away from his extraordinary achievements. Winner of the German F3 title after only two attempts, he graduated to Formula One in 1991, made his debut for Jordan but quickly settled with Benetton and took his first world title in 1994, defending it successfully the following year.

Taking on next the enormous challenge of reviving Ferrari, he established himself as one of the greats, clinching his third and Ferrari's first world title for 21 years in 2000, sparking a run of five consecutive titles. The streak ended in 2005, finally dethroned by Renault's Fernando Alonso, yet he might have won again in 2006 but for suffering his first engine failure in six years while leading in the penultimate race, by which time he had already announced his retirement. His comeback has not brought new success, the flashes of breathtaking genius of his earlier career no longer there. Ninth in 2010 and 8th in 2011, his 3rd place at the European GP in 2012 was his first podium finish since his final win for Ferrari. Yet nothing can alter what went before.

BORN: June 27, 1985, Wiesbaden, Germany

2012 TEAM: Mercedes

DEBUT: 2006 Bahrain GP (Williams)

FIRST FORMULA ONE WIN: 2012 Chinese GP (Mercedes)

BEST FORMULA ONE SEASON: 7th in 2009 (Williams), 2010 and 2011 (both Mercedes)

NICO ROSBERG

ROSBERG ONCE TOYED WITH THE IDEA OF A CAREER IN TENNIS, BUT AS THE SON OF THE FINNISH 1982 FORMULA ONE WORLD CHAMPION, KEKE, IT WAS ALMOST INEVITABLE HE WOULD END UP BEHIND THE WHEEL OF A RACING CAR.

His talent was obvious. After five years at the top in karting, he entered German Formula BMW and was champion at the first attempt, aged just 17. As the youngest person to drive a Formula One car, he tested for BMW Williams team in 2002, and again in 2003 and 2004, before joining the ART squad for the inaugural GP2 series, which he also won.

Offered a Formula One race seat by Williams for 2006, he finished 7th on his debut and scored the bulk of the team's points in 2007, before improving still more to take a third and a second in 2008, despite a generally uncompetitive car.

The podium eluded him in 2009, after which he signed with double champions Mercedes for 2010. Rule changes then robbed his new team of their domination and in two years three third places was the extent of his success. His fortunes changed dramatically at the 2012 Chinese Grand Prix, however, when Rosberg not only took his first pole but stormed to victory by a 20-second margin, securing his first race win, a first for Mercedes since Juan Manuel Fangio in 1955 and the first in Formula One history by a German driving a German car.

KIMI RÄIKKÖNEN

KIMI RÄIKKÖNEN HAD DRIVEN ONLY 23 SINGLE-SEATER RACES WHEN SAUBER GAVE HIM A FORMULA ONE TEST.

The former karting star had won seven races to be UK Formula Renault champion in 2000. Sauber were so impressed they handed Räikkönen a race drive at the 2001 Australian GP. He ended his first season with nine championship points, having driven with pace and consistency.

Soon McLaren had him earmarked as a successor to two-times champion, Mika Häkkinen, the original 'flying Finn'. There were fears it might be a step too far for an inexperienced driver but Räikkönen silenced the skeptics, notching four podium finishes and 24 points. The 2003 season brought his first victory, in Malaysia, followed by 10 more podiums, giving him enough points to run Michael Schumacher close in the championship. He was runner-up again in 2005 – unluckily perhaps – when he won seven times but could not overhaul the Renault of Fernando Alonso. It paved the way for a move to Ferrari in 2007 and his first drivers' title.

Replacing Michael Schumacher was a tough challenge but Räikkönen won on his team debut in Australia and gained further victories in France, Britain, Belgium and China before regaining the lead in the final race in Brazil. After two seasons that were disappointing by comparison, he quit Formula One, spending three seasons in the World Rally Championship. He returned to Formula One with Lotus in 2012.

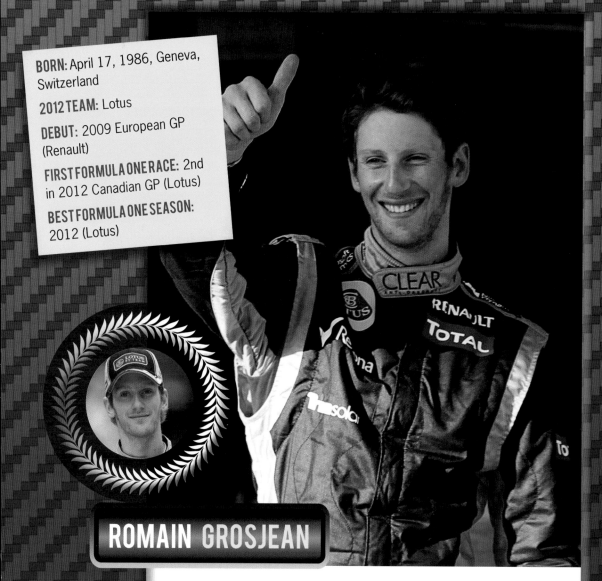

BORN: April 17, 1986, Geneva, Switzerland

2012 TEAM: Lotus

DEBUT: 2009 European GP (Renault)

FIRST FORMULA ONE RACE: 2nd in 2012 Canadian GP (Lotus)

BEST FORMULA ONE SEASON: 2012 (Lotus)

ROMAIN GROSJEAN

ALTHOUGH BORN IN GENEVA, GROSJEAN RACES UNDER THE FRENCH FLAG. THE 2012 SEASON, WHICH BROUGHT HIS FIRST PODIUM FINISHES, WAS HIS THIRD IN FORMULA ONE, HAVING FOLLOWED A TRADITIONAL ROUTE UP THE RANKS.

Swiss Formula Renault 1600 champion in 2003, he took the French Formula Renault title in 2005 and, racing for ASM, won the Formula Three Euro Series at the second attempt after winning six races in 2007, by which point he had secured a place on Renault's driver development programme. He graduated to GP2 in 2008, winning the inaugural GP2 Asia Series with ART and finishing 4th in the main GP2 Series, while also testing in Formula One for the first time with Renault.

Installed as third driver for Renault, he was promoted to an Formula One race drive sooner than he expected in 2009 following the cancellation of Nelson Piquet Jnr's contract but failed to impress in seven races and was dropped for the 2010 campaign.

Seeking to rebuild his confidence, he competed away from Formula One, winning the Auto GP championship and two FIA GT events, while also appearing in the Le Mans and Spa 24-hour races. Back in GP2 in 2011, he excelled again, winning both the Asian and the main series after scoring six victories. It was enough to earn another chance with Renault in Formula One in 2012, to which he responded with third place in Bahrain and second in Canada.

05 CHINA

VENUE: Shanghai International Circuit

CIRCUIT LENGTHS: 5.451km

TURNS: 16

LAPS: 56

06 BAHRAIN

VENUE: Bahrain International Circuit, Sakhir

CIRCUIT LENGTHS: 5.412km

TURNS: 15

LAPS: 57

COSTING AN ESTIMATED $450M WHEN IT WAS COMPLETED IN 2004, THE SHANGHAI INTERNATIONAL CIRCUIT WILL HOST ITS 10TH FORMULA ONE GRAND PRIX IN 2013 AND HAS BEEN WORTH EVERY PENNY OF AN INVESTMENT THAT SAW AN AREA OF SWAMPLAND TRANSFORMED INTO A STUNNING RACETRACK.

Shanghai was the final venue of the season in 2005 and world champion Fernando Alonso's victory in China that year secured Renault the Constructor's Championship.

McLaren's Lewis Hamilton could possibly claim to have the fondest memories of the circuit as the only multiple winner in Shanghai; his 2008 triumph left him on the cusp of his maiden world title while his success in 2011 provided him welcome relief in a season otherwise largely dominated by Sebastian Vettel.

The Chinese race has been a commercial success but the annual event, held in the Jiading district of Shanghai, has so far not resulted in a Chinese driver competing in Formula One, although Ma Qing Hua, incorporated into HRT's driver development programme last year, could end that wait in the near future.

DID YOU KNOW?

The very long back straight at Shanghai is the largest in Formula One, at 1170m, which is roughly the equivalent of 11 football pitches.

OFF-TRACK CONTROVERSIES HAVE FAILED TO DETRACT FROM THE EXCITING RACING SEEN IN BAHRAIN SINCE FORMULA ONE FIRST CAME TO THE COUNTRY IN 2004.

The remote desert location was not used in 2011 because of political demonstrations in Bahrain, but one man who will certainly be keen for the venue to keep its position in the calendar is Fernando Alonso. The current Ferrari driver has won three of the eight races staged in Bahrain.

The 15-turn circuit provides three genuine overtaking opportunities, partly due to the variation in the track's width. There are very wide sections because the venue was built with flexibility in mind. There are five track layouts within the same grid, the Formula One track being accompanied by an endurance circuit, a testing oval and a drag strip amongst others.

Ferrari have won four times in Bahrain, but it was Red Bull's Sebastian Vettel who won there last year. The reigning champion secured his first victory of the 2012 season in Sakhir.

DID YOU KNOW?

Alcohol is legal in Bahrain but champagne is not sprayed by the winners on the podium, as is the norm at most races. Instead, a non-alcoholic rosewater drink known as Waard is used.

07 SPAIN

VENUE: Circuit de Catalunya, Barcelona

CIRCUIT LENGTHS: 4.655km

TURNS: 16

LAPS: 66

BARCELONA IS A FAMILIAR VENUE FOR FORMULA ONE DRIVERS, HAVING HOSTED THE SPANISH GRAND PRIX RACE EVERY YEAR SINCE 1991. THE CATALUNYA CIRCUIT IS ALSO REGULARLY USED BY TEAMS TO TEST THEIR CARS.

Built with the 1992 Olympic Games in mind, the track is known as one suiting an all-round driver; it has a combination of challenging corners and long, fast straights. Overtaking is not considered easy, with Elf corner thought to be the most sensible point to attempt a manoeuvre. Indeed, the 1999 race, won by McLaren's Mika Häkkinen, is noteworthy as there was only one reported overtaking move in the race.

Britain's Nigel Mansell won the first two races in Barcelona but it is a German who has experienced the most success at the Catalunya track. Michael Schumacher has six Spanish Grand Prix victories to his name, including four consecutively between 2001 and 2004.

Jenson Button was the last Brit to triumph at the Circuit de Catalunya, while last year's race was surprisingly won by Venezuelan Pastor Maldonado, who in the process became the first man from his country to appear on a Formula One podium, let alone win a race.

DID YOU KNOW?

The Spanish Grand Prix has been involved in Formula One since 1951, but Fernando Alonso's victory in Barcelona in 2006 was the first time a Spaniard won the race.

08 MONACO

VENUE: Circuit de Monaco, Monte Carlo

CIRCUIT LENGTHS: 3.340km

TURNS: 19

LAPS: 76

FORMULA ONE'S STANDOUT RACE, THE MONACO GRAND PRIX IS THE EVENT ALL BUDDING RACING DRIVERS DREAM OF WINNING. THE CHARMING URBAN TRACK IN MONACO HAS BARELY CHANGED SINCE BECOMING AN ANNUAL FEATURE OF FORMULA ONE IN 1955.

The circuit requires superb car control and significant bravery from drivers if they are to be successful, with the narrow track making overtaking extremely difficult, so qualifying high up the grid is especially important at Monaco.

The great Ayrton Senna made the Circuit de Monaco his own, winning six times in total – the most of any driver including five in a row between 1989 and 1993. British drivers have also generally fared well in Monte Carlo; Jenson Button and Lewis Hamilton have both already won the event, Stirling Moss and Jackie Stewart have three Monaco wins apiece while Graham Hill won five times during the 1960s.

The glamorous principality on the Mediterranean coast regularly attracts A-list celebrities on a race weekend with recent visitors including Brad Pitt and Liz Hurley.

DID YOU KNOW?

Monaco has never produced a race-winning Formula One driver but because of its tax status it has been home to many of the sport's leading lights over the years, including Ayrton Senna, Gilles Villeneuve, David Coulthard and Jenson Button.

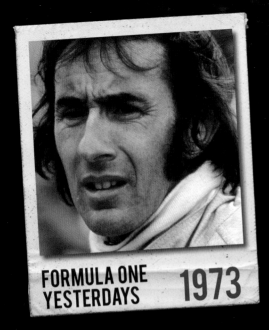

FORMULA ONE YESTERDAYS **1973**

1973

STEWART'S YEAR OF TRIUMPH AND TRAGEDY

Jackie Stewart had already decided he would retire when the 24th season of Formula One began 40 years ago. The fifth British driver – and the second Scot – to become World Champion, Stewart was by then 33 and had already taken the drivers' title twice during his legendary partnership with Ken Tyrrell.

By now the Formula One year consisted of 15 championship races, beginning in Argentina in January and ending in the United States in October. In the German Grand Prix in August, Stewart won his fifth race of the year in the Elf-sponsored Tyrrell 005 and his 27th in total, setting a record for career wins that would stand for 14 years.

A month later, fourth place in the Italian GP ensured he would reclaim the title taken from him a year earlier by the Brazilian, Emerson Fittipaldi, and thus go out at the top, although his plan to call time on his career was still a secret he had shared only with Tyrrell. He would reveal it after the US race, announcing that the Frenchman Francois Cevert, who had joined Tyrrell on Stewart's recommendation in 1970, would succeed him as number one driver in 1974.

But in 1973 accidents were still common and tragedy never far away. In practice at the Watkins Glen circuit, while negotiating the tricky uphill section known as The Esses, Cevert lost control after clipping a kerb and was killed when his car collided with a guardrail. Stewart and Tyrrell immediately withdrew from the race.

```
T S O R P H M A H B A R B L M
V I R T V I L L E N E U V E W
D D W Z H M A S T N R N Q N M
D L V J M U E C G Q Z L Y Z L
X A E R D N T M T R A W E T S
P P T A N R H A M I L T O N R
I I T A N E L F N L M M F L
Q T E Y E H G H M B X J O R R
U T L R N C L Y O A U S M L M
E I N L I A M Z F I N T H D R
T F L K K M C V D O G S T K M
C I D T K U W J L M V N E O X
H P N B A H H A V M F L A L N
V D R Z H C S E E T R U S F L
K N J N G S W Z G J G K K K T
```

ALONSO
BRABHAM
BUTTON
FANGIO
FITTIPALDI
HAKKINEN
HAMILTON
HILL
LAUDA
MANSELL
PIQUET
PROST
SCHUMACHER
SENNA
STEWART
SURTEES
VETTEL
VILLENEUVE

CHAMPIONS WORDSEARCH

Find the names of 18 World Champion Drivers
hidden in this grid. Words can go horizontally,
vertically and diagonally in all eight directions.
Answers on page 62.

BORN: September 13, 1986; Hyogo, Japan

2012 TEAM: Sauber

DEBUT: 2009 Brazilian GP (Toyota)

BEST FORMULA ONE RACE: 5th in 2011 Monaco GP and 2012 Spanish GP (both Sauber)

BEST FORMULA ONE SEASON: 12th in 2010 (BMW Sauber) and 2011 (Sauber)

KAMUI KOBAYASHI

KOBAYASHI, WHOSE FATHER OWNS A SUSHI RESTAURANT, WON A STRING OF JUNIOR KARTING TITLES IN JAPAN AND LANDED A PLACE ON TOYOTA'S COVETED YOUNG DRIVERS' PROGRAMME IN 2001.

In 2003 he finished second overall in Formula Toyota and, with the company's financial backing, sought further opportunities in Europe. A creditable 7th in the Italian Formula Renault series in 2004, he contested both the Italian and Euro series in 2005 and won both.

Stepping up to Formula Three, he finished 8th in the Euro series in 2006 and 4th in 2007, in the meantime testing for Toyota in Formula One, impressing enough to be signed up as their third driver for 2008. At the same time, he raced in the GP2 Asia and main series, winning the Asia series in 2009 for the DAMS team, although less successful in the main series, finishing 16th in 2008 and 2009. Kobayashi's first Formula One races came late in the 2009 season after Timo Glock suffered a cracked vertebra.

He narrowly missed out on a point on his debut at the Brazilian before powering into 6th place in Abu Dhabi. Toyota then withdrew from Formula One but Kobayashi had done enough to find himself sought after and he signed to race full-time for BMW Sauber in 2010. An aggressive, exciting driver, he finished 12th in 2010 and 2011, finishing 5th in Monaco in his best drive.

SERGIO PEREZ

BORN: January 26, 1990; Guadalajara, Mexico

2012 TEAM: Sauber

DEBUT: 2011 Bahrain GP (Sauber)

FIRST FORMULA ONE RACE: 2nd in 2012 Malaysian GP (Sauber)

BEST FORMULA ONE SEASON: 16th in 2011 (Sauber)

WHEN SERGIO PERÉZ MADE HIS DEBUT IN 2011, HE BECAME ONLY THE FIFTH MEXICAN TO COMPETE IN FORMULA ONE AND THE FIRST SINCE HÉCTOR REBAQUE RETIRED IN 1981.

Another with a successful background in karting, Peréz made his single-seater debut in the 2004 Skip Barber National Championship in the United States, with financial backing from Escuderia Telmex, which supports racing drivers.

Moving to Europe, he finished 14th in Germany's Formula BMW ADAC series in 2005 and 6th in 2006, before stepping up to the national class of British Formula Three, winning his first single-seater title and promotion to international class for 2008, when he achieved four wins in finishing 4th overall.

The following year, in the Asia GP2 series, he finished 7th before enjoying a hugely successful 2010 campaign in the main GP2 series, driving for Barwa Addax, collecting five wins and seven podiums to finish runner-up to Pastor Maldonado. It was enough to win him a Formula One seat with Sauber for 2011 before he had even tested but Peréz proved himself immediately, coming home 7th on his Bahrain debut.

A crash in qualifying in Monaco set him back, costing him two race appearances and undermining his confidence, but he still finished 16th and rewarded Sauber's faith in him again in the 2012 Malaysian GP, finishing 2nd to Fernando Alonso from 9th on the grid. It was Sauber's first podium finish.

DRIVER PROFILES 10

DANIEL RICCIARDO

DANIEL RICCIARDO WAS INSPIRED BY AYRTON SENNA TO TAKE UP MOTOR RACING, AND MADE HIS SINGLE-SEATER DEBUT IN AUSTRALIAN FORMULA FORD IN 2005.

He did well enough in a 15-year-old Van Diemen to earn a scholarship for the following season's Formula BMW Asia championship, in which he finished 3rd with the Eurasia team.

In 2007 he raced in Italian Formula Renault, ending the season in 6th place, before landing his first title the following year in the Formula Renault Western European Cup with eight wins from 15 races. He also finished runner-up in the European series. In 2009 he made another advance, winning the British Formula Three title.

He was by this time a member of the Red Bull junior team and, after finishing 2nd in the Formula Renault 3.5 series in 2010, he became a regular third driver for the Red Bull-owned Toro Rosso team. His performances in practice gave rise to talk of a race seat. In the event, he continued to compete in Formula Renault 3.5 and, in an arrangement brokered by Red Bull to further his development, he made his Formula One debut in the 2011 British Grand Prix with HRT, replacing Narain Karthikeyan, and remained with that team for the rest of the season.

Although he did not pick up any points, he made a good enough impression to secure a drive with Toro Rosso for 2012.

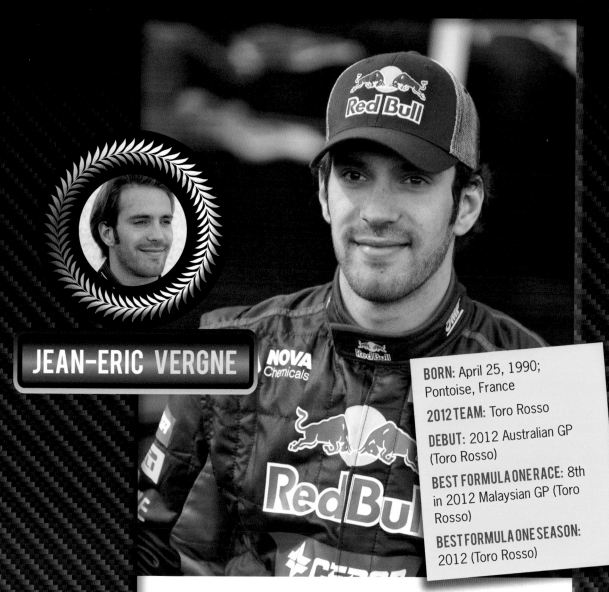

JEAN-ERIC VERGNE

BORN: April 25, 1990; Pontoise, France

2012 TEAM: Toro Rosso

DEBUT: 2012 Australian GP (Toro Rosso)

BEST FORMULA ONE RACE: 8th in 2012 Malaysian GP (Toro Rosso)

BEST FORMULA ONE SEASON: 2012 (Toro Rosso)

BORN IN THE PARIS SUBURB OF PONTOISE IN 1990, JEAN-ERIC VERGNE GRADUATED FROM KARTING TO SINGLE-SEATER CAR RACING IN 2007, JOINING THE FRENCH FORMULA CAMPUS SERIES AND WINNING THE TITLE WITH SIX VICTORIES.

He was soon invited to join Red Bull's young driver programme, which gave him the platform to compete in the both the Eurocup Formula Renault 2.0 and West European Cup in 2008. He took the French Formula Renault title that year as the best French driver in the WEC standings.

In 2009 he was runner-up in both series before making a highly successful transition to British Formula Three, winning the prestigious title at the first attempt in 2010 with an impressive 13 victories. He made his Formula One test debut with Toro Rosso in November of the same year and also took part in four GP3 races. Back in Europe in 2011, he finished runner-up in the Formula Renault 3.5 series and drove for Toro Rosso during first free practice at three of the last four Grands Prix.

After impressing in Formula One's young driver test after the Abu Dhabi GP in November of the same year, with a lap time only half a second off Sebastian Vettel's pole time, Vergne landed a Formula One race drive alongside Daniel Ricciardo for the 2012 season with Toro Rosso, hailed potentially as the latest in an illustrious line of British F3 champions to become a star in Formula One.

09 CANADA

VENUE: Circuit Gilles Villeneuve, Montreal

CIRCUIT LENGTHS: 4.361km

TURNS: 14

LAPS: 70

RENAMED AFTER THE GREAT GILLES VILLENEUVE AFTER HIS TRAGIC DEATH IN 1982, THE MONTREAL CIRCUIT HAS BEEN A REGULAR IN FORMULA ONE SINCE IT WAS BUILT IN THE LATE 1970S.

For years the French and English parts of Canada alternated Formula One races between them but when the Mont-Tremblant track in the French region was deemed too dangerous, French-Canadians decided to connect all the roads on the Ile Notre-Dame to make a circuit. The developed track has staged the Canadian Grand Prix every year since, apart from in 2009 when the event was not held.

Fittingly, it was Villeneuve who won the first Grand Prix held at the new circuit in 1978 but he remains the only Canadian to have won the event at the track. Michael Schumacher has seven victories at Montreal the most of any driver – including six for Ferrari, who are the constructor with the most wins at the Notre-Dame circuit, having won there ten times.

Four of the last five Formula One races in Montreal have been won by British drivers: Lewis Hamilton has three wins in 2007, 2010 and 2012 while Jenson Button won for McLaren in 2011.

In 2009 a deal was agreed with the FIA for the Canadian Grand Prix to be staged at the Gilles Villeneuve Circuit until 2014.

DID YOU KNOW?

The final corner of the Montreal circuit has become notorious for crashes involving World Champions. In 1999, Damon Hill, Michael Schumacher and Jacques Villeneuve all collided with the wall on that corner, which has since become known as the Wall of Champions.

10 GREAT BRITAIN

VENUE: Silverstone Circuit, Northamptonshire

CIRCUIT LENGTHS: 5.891km

TURNS: 18

LAPS: 52

STRADDLING THE BORDER BETWEEN NORTHAMPTONSHIRE AND BUCKINGHAMSHIRE, SILVERSTONE HAS BEEN A REGULAR HOST OF THE BRITISH GRAND PRIX SINCE 1948.

Unlike other courses with a rich and long history of staging racing, the English circuit has hardly altered in its appearance since it made its Formula One debut in 1950.

Silverstone has staged the British weekend in Formula One every year since 1987 but from 1955 to 1962 it alternated with Aintree before the Merseyside venue fell out of favour in the 1960s. Silverstone then had to share the right to stage Formula One races with Brands Hatch in Kent until the mid 1980s.

French legend Alain Prost has the most Silverstone victories, winning in 1983, 1985, 1989, 1990 and 1993. A host of British drivers have won at the rural track – David Coulthard, Nigel Mansell and Jackie Stewart amongst them with Lewis Hamilton the most recent home winner, triumphing at Silverstone in his Championship winning year of 2008.

In 2008 a tentative agreement was reached for Donington Park to become the host circuit for the British race weekend from 2010 but the rights were returned to Silverstone when the East Midlands circuit failed to secure the necessary funding.

DID YOU KNOW?

The first British Formula One winner at Silverstone was the Ferrari driver Peter Collins. It was his third race victory from 35 starts. Tragically, in his very next race, he was killed after a crash in the German Grand Prix at the Nürburgring.

11 HUNGARY

VENUE: Hungaroring, Budapest

CIRCUIT LENGTHS: 4.381km

TURNS: 14

LAPS: 70

THE FIRST VENUE BEHIND THE IRON CURTAIN TO HOST FORMULA ONE RACING, THE HUNGARORING IN BUDAPEST HAS NOW HOSTED 27 GRAND PRIX WEEKENDS SINCE IT WAS FIRST INCLUDED IN THE CALENDAR IN 1986.

Located 19km from the centre of Budapest, the track was built only after a proposed street circuit in the middle of the city was deemed impractical.

Michael Schumacher has won the most races in Hungary with four while Brazilian legends Ayrton Senna (3) and Nelson Piquet (2) are also multiple winners in Budapest. Former World Champions Fernando Alonso and Jenson Button both secured their first race wins in Budapest, in 2003 and 2006 respectively.

Its narrow and twisty nature has made overtaking difficult traditionally at the Budapest circuit, although attempts to remedy that were made in 2003.

There have still been some epic races at the circuit, however, the most notable probably coming in 2007, when Damon Hill sensationally passed Michael Schumacher's Ferrari on the last lap only for a mechanical failure to consign him to second place.

DID YOU KNOW?

Formula One chiefs wanted Moscow, in the Soviet Union, to be the first Grand Prix venue behind the Iron Curtain but a deal was never reached. In 1983 their attentions turned to Hungary and Moscow's loss has been Budapest's gain.

12 BELGIUM

VENUE: Circuit de Spa-Franchorchamps

CIRCUIT LENGTHS: 7.004km

TURNS: 19

LAPS: 44

FAST, HILLY AND TWISTY, SPA IN BELGIUM HAS ESTABLISHED ITSELF AS ONE OF THE MOST CHALLENGING AND COMPETITIVE RACES IN THE FORMULA ONE CALENDAR, A FAVOURITE OF MOST DRIVERS.

The track, situated between the Belgian towns of Spa, Malmedy and Stavelot, first hosted the Belgian Grand Prix in 1925 and was one of the venues for Formula One's inaugural Championship year in 1950.

The circuit was at one point almost twice the length of its current configuration, but many of its famous turns and features were present when the first races were held at Spa in 1922. One particularly challenging aspect of the track is the renowned Eau-Rouge 'left-right-left' sequence of corners, which launches drivers steeply uphill before arriving at a blind summit. To compound this drivers must attempt Eau-Rouge flat out primarily to maximise downforce but also to carry as much speed as possible into the uphill straight which follows the turn.

German legend Michael Schumacher boasts six Spa wins the most of any driver while Briton Jim Clark won four times at the track in the 1960s. Ayrton Senna, meanwhile, won four times consecutively between 1988 and 1991.

DID YOU KNOW?

The Belgian Grand Prix left Spa-Francorchamps between 1972 and 1982 circuits at Zolder and Nivelles both hosting it instead during this period but the event returned to Spa permanently in 1984.

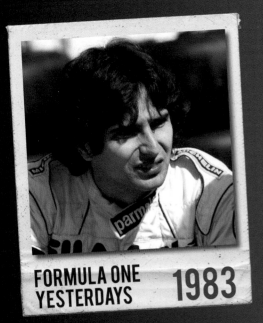

FORMULA ONE YESTERDAYS 1983

PIQUET HANGS ON TO DENY PROST

The British Grand Prix of 1983 – the last to be staged on a Saturday – was won driving by Alain Prost, who was bidding to become the first Frenchman to win the Formula One drivers' title and who now had a six-point lead over Brazil's Nelson Piquet in the standings. When he won in Austria two rounds later his advantage stood at 14 points.

But for Prost the next two rounds were calamitous. Trying to overtake Piquet in Holland, he collided with the Brazilian's Brabham and spun off. Next, in Italy, the turbo on Prost's Renault failed and again he took no points. Piquet, meanwhile, won both at Monza and in European GP, this time staged at Brands Hatch, and although Prost battled through from 8th on the grid to finish runner-up at the Kent circuit, his lead going into the last of the 15 races, in South Africa, was down to just two points.

With Piquet breathing down his neck, Prost knew he had to have a good race at Kyalami, so it did not help that he was only fifth quickest in practice.

He clawed his way up to third place in the race but on lap 35 the RE40's turbo failed again and his race was over. Piquet had turbo problems with the Brabham too and lost power, allowing teammate Riccardo Patrese to go through for his first GP win, but the points he took for finishing 4th were enough to pip Prost by two points.

20 QUESTIONS QUIZ

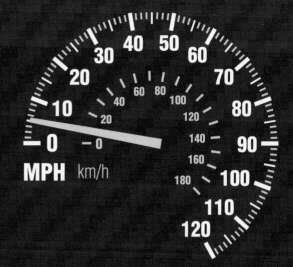

1 Which Formula One world champion once played in the same school football team as England star Ashley Young?

2 Which iconic track is the shortest on the Formula One circuit?

3 Which legendary British driver finished second in the drivers' world championship four times but never won the title?

4 Which Formula One venue includes Stowe, Club and Becketts corners?

5 Three drivers with the surname Hill have won the world championship. Graham Hill and son Damon Hill are two: who is the third?

6 How many Grand Prix races had Britain's Jenson Button won before his 2009 world title season?

7 Which future British world champion driver made his Grand Prix debut at Monaco in 1973?

8 Since Formula One was launched in 1950, which two circuits other than Silverstone have hosted the British Grand Prix?

9 Which Italian driver started 208 Grand Prix races between 1980 and 1994 but finished his career without a win?

10 Which future world champion retired on the first lap of his Formula One debut at the 1991 Belgian Grand Prix?

11 What was special about Pastor Maldonado's victory at the 2012 Spanish Grand Prix?

12 Which world champion is the only Spanish-born driver to win in Formula One?

13 Jackie Stewart's record of 27 Grand Prix wins was the most by a British driver until Nigel Mansell won at Silverstone in 1992: how many GPs did Mansell win in total?

14 Which Italian-born American became world champion driver in 1978?

15 Which British driver won the Monaco Grand Prix twice in three years between 2000 and 2002?

16 Which British world champion driver tried to qualify for the Great Britain trap shooting team at the 1960 Olympics?

17 Legendary team boss Ron Dennis is associated with which iconic Formula One name?

18 Which British world champion driver was born in Frome, Somerset?

19 Why is the small northern Italian town of Maranello so famous in the history of Formula One?

20 Since the late Ayrton Senna took the last world championship won by a Brazilian in 1991, which two Brazilians have finished second?

13 ITALY

VENUE: Autodromo Nazionale Monza, Monza

CIRCUIT LENGTHS: 5.793km

TURNS: 11

LAPS: 53

14 SINGAPORE

VENUE: Marina Bay Street Circuit, Singapore

CIRCUIT LENGTHS: 5.073km

TURNS: 23

LAPS: 61

THE HOME OF FORMULA ONE IN ITALY, A COUNTRY SYNONYMOUS WITH FERRARI AND MOTORSPORT, MONZA HAS HOSTED A WEEKEND IN THE CALENDAR EVERY YEAR BUT ONE WHEN IMOLA HOSTED THE ITALIAN EVENT IN 1980 SINCE THE WORLD CHAMPIONSHIP BEGAN IN 1950.

The track has scarcely altered in appearance since the 1950s, although a full circuit which included the testing oval was used in 1955, 1956, 1960 and 1961.

Monza is known for being an extremely fast track, with long straights and high speed corners, and the minor modifications that have been made over the years have usually been attempts to reduce turn speeds.

Home to some of the best races and regrettably some of the worst accidents Monza to many embodies the compelling appeal of Formula One; hard, fast, uncompromising racing. The Italians call it 'La Pista Magica' (the magic track), a grand description which appears fully justified.

Ferrari have been dominant on home turf, with around twice as many race wins at Monza than any of their rivals, while Michael Schumacher's five race wins at Monza is an unrivalled figure.

DID YOU KNOW?

Despite Italy's rich tradition in motor racing, the last Italian to win at Monza was way back in 1966, when Ludovico Scarfiotti won for Ferrari.

A RECENT ADDITION TO THE FORMULA ONE CALENDAR, THE SINGAPORE GRAND PRIX HAS PROVED A HUGE SUCCESS SINCE THE COUNTRY HOSTED ITS FIRST RACE IN 2008.

That Grand Prix was the first in Formula One history to be staged at night, with the city-state's famous and spectacular skyline providing a stunning, illuminated backdrop for a Fernando Alonso victory.

The circuit is partly staged on public roads and organisers use powerful lighting systems to replicate daylight conditions for the drivers. The timing of the event not only attaches a novelty factor to the Singapore weekend, it also means the race takes place at a convenient time for European TV audiences.

Interesting features include a stretch of the circuit that runs over the Anderson Bridge as well as the section between turns 18 and 19 which goes underneath the Grandstand.

With heat and humidity the prevailing weather conditions in Singapore, the race has also become one of Formula One's most unforgiving, with the numerous bumps and kerbs exerting physical pressure on the drivers.

DID YOU KNOW?

Singapore's government envisaged the race as part of a national festival when they first won the rights to stage Formula One in 2008. Their dream seemed to have been realised in 2010 as the race took place during 10 days of parties, concerts and exhibitions.

15 JAPAN

VENUE: Suzuka Circuit

CIRCUIT LENGTHS: 5.807km

TURNS: 18

LAPS: 53

AFTER BEING BUILT BY HONDA AS A TEST FACILITY IN 1962, SUZUKA HAD TO BE PATIENT IN WAITING TO HOST THE JAPANESE GRAND PRIX, WITH THE FUJI SPEEDWAY PREFERRED FOR MANY YEARS.

In 1978, the Japanese weekend was scratched from the calendar. When it was reinstated, in 1987, the event was moved to Suzuka, where it has been staged every year since, apart from 2007 and 2008.

The circuit's figure of eight layout is unusual and the track boasts some of the most challenging corners in Formula One, most notably the 130R and the Spoon Curve.

While slightly earlier in the season now, the Japanese Grand Prix has traditionally been very late in the calendar sometimes last so many World Champions have been crowned at Suzuka. Michael Schumacher whose six race wins make him the most successful driver at Suzuka sealed his 2000 and 2003 crowns in Japan.

A decade earlier, Suzuka played host to a famous spat between Alain Prost and Ayrton Senna. In 1989 the pair – both driving for McLaren – collided, forcing Prost to retire. Senna went on to win the race but was disqualified for his role in the incident with Prost, who sealed the World Championship as a result.

DID YOU KNOW?

Current Sauber driver Kamui Kobayashi is the 20th Japanese driver in Formula One but none of the previous 19 has ever won a Grand Prix.

16 SOUTH KOREA

VENUE: Korea International Circuit, Yeongam

CIRCUIT LENGTHS: 5.615km

TURNS: 18

LAPS: 55

IN A STUNNING WATERSIDE LOCATION, THE KOREAN GRAND PRIX HAS BEEN AN INSTANT HIT SINCE ITS INAUGURATION IN 2010.

Made up by a permanent track combined with a temporary street circuit, the Hermann Tilke-designed venue features an enormous 1.2km straight as well as several turns, some slow and some fast.

The largely rural venue is 400km from Seoul but the city of Jeollanam-do is expected to grow around the circuit, with plans for business parks and entertainment complexes already developed.

Korean promoters agreed a seven-year deal for the event with Bernie Ecclestone in 2006, with a five year extension included as an option, meaning the race should form part of the Formula One calendar until at least 2021.

In 2010, Ferrari's Fernando Alonso claimed victory in the maiden Korean Grand Prix while Sebastian Vettel's 2011 triumph sealed the Constructors' Championship for Red Bull with three races to spare. The previous day, McLaren's Lewis Hamilton had broken Red Bull's incredible streak of 15 consecutive pole positions, which dated back to the 2010 Brazilian Grand Prix.

DID YOU KNOW?

The 16,000 seater grandstand in Korea features a roof which has been designed to resemble the distinctive eaves of Korea's traditional 'Hanok' houses.

BORN: January 19, 1980; Frome, England

2012 TEAM: McLaren

DEBUT: 2000 Australian GP (Williams)

FIRST FORMULA ONE WIN: 2006 Hungarian GP (Honda)

FIRST FORMULA ONE TITLE: 2009 (Brawn GP)

DRIVER PROFILES **13**

JENSON BUTTON

JENSON BUTTON WENT A LONG WAY TO PROVING THAT HE SIGNED FOR MCLAREN TO BE FAR MORE THAN LEWIS HAMILTON'S UNDERSTUDY BY OUTPERFORMING HIS TEAMMATE IN 2011.

The 2009 world drivers' champion surprised many by teaming up with his fellow Brit but proved doubters wrong as Hamilton finished behind a teammate in the standings for the first time.

Button, who turns 33 in 2013, made his debut in Formula One in 2000 but had to endure a frustrating six-year wait for his maiden Grand Prix victory.

In his early career, Button knew only success. He won karting competitions in Britain and Europe and took the British Formula Ford title at the first attempt in 1998.

The following year he claimed three victories en route to a third placed finish in British Formula Three, a performance which earned him the competition's Rookie of the Year award and a Formula One berth with Williams.

Button's fortunes at the top level were mixed. He drove for Williams, Renault, BAR and Honda in his first nine seasons in Formula One but his first race win – in the 2006 Hungarian Grand Prix for Honda – was his only victory in that period, although he did finish third in the 2004 drivers' championship with BAR, when he had ten podium finishes.

But his luck changed after Honda withdrew from Formula One in 2008. A late management buyout from Ross Brawn saved the team and secured Button a place on the grid for 2009, when Brawn GP's innovative diffuser design saw the Somerset-born driver make a blistering start. He won six of the first seven races and while Brawn's pace advantage was eventually cut by their rivals, Button clinched his maiden world championship in Brazil with a race to spare.

After moving to McLaren, Button delivered their first two race wins of 2010 but the team could not match Red Bull and Ferrari for pace and the reigning champion had to settle for fifth place, one behind teammate Hamilton. It was always a risk when Button joined McLaren that he would be overshadowed by Hamilton's precocious talent.

In 2011, however, Button silenced doubters by winning three races in a season dominated by eventual champion Sebastian Vettel. He finished second in the drivers' standings, three places ahead of his teammate.

DID YOU KNOW?

Button, aged 17, needed only six driving lessons to pass his test although it was at the second attempt. He failed the first time for driving too close to a parked car on a narrow street.

BORN: January 7, 1985;
Stevenage, England

2012 TEAM: McLaren

DEBUT: 2007 Australian GP
(McLaren)

FIRST FORMULA ONE WIN: 2007
Canadian GP (McLaren)

FIRST FORMULA ONE TITLE: 2008
(McLaren)

LEWIS HAMILTON

STEVENAGE-BORN LEWIS HAMILTON HAS ALWAYS BEEN A WINNER. HE WAS THE YOUNGEST EVER WINNER OF THE BRITISH CADET KART CHAMPIONSHIP – AGED ONLY 10 – AND WON BOTH THE FORMULA THREE EUROSERIES AND GP2 CHAMPIONSHIPS BEFORE HE MADE THE STEP UP TO FORMULA ONE.

Few were surprised therefore when, in only his second year at the top level in 2008, Hamilton became the youngest ever Formula One world champion, a record subsequently broken by Sebastian Vettel.

Hamilton, who was first signed by McLaren to their young driver programme in 1998, excelled in Formula One after surprisingly being picked by the British team to partner Fernando Alonso in 2007. He led the race for much of the season, picking up four Grand Prix victories along the way, and was eventually beaten to the title – won by Ferrari's Kimi Räikkönen – by only one point. This performance, coupled with the way he matched team-mate and world champion Alonso, boded well for the following year and Hamilton duly delivered his maiden world title.

Then 23, Hamilton again led the championship for much of the season, eventually clinching the title on the final weekend in Brazil. He famously sealed the fifth place finish he required to pip Felipe Massa to the title on the final corner of the final lap.

His astonishing title triumph was rewarded by the Queen with an MBE in the 2009 New Year honours.

McLaren have struggled since to compete with the sport's front runners for pace but Hamilton's attacking style and race instincts have continued to attract praise.

He finished well short of champion Jenson Button in 2009 but competed well the following year, his title challenge ending only at the final race.

Hamilton, who recently moved to Monaco after living in Switzerland for many years, finished below his teammate Button in the drivers' championship in 2011 for the first time in his career. A host of uncharacteristic errors and accidents, usually involving Ferrari's Massa, hamstrung him and he eventually finished only fifth.

The 28-year-old remains one of the most exciting drivers on the circuit, however, and is renowned for his aggressive driving style and confident car control. He thrives in wet conditions, typified by his 2008 victory at Silverstone where he won in heavy rain after qualifying fourth, as he is highly proficient at overtaking, his aggression often forcing errors from his rivals.

DID YOU KNOW?

Lewis raised £42,100 for the charity Tommy's, which funds research into pregnancy problems, by selling one of his karts on eBay.

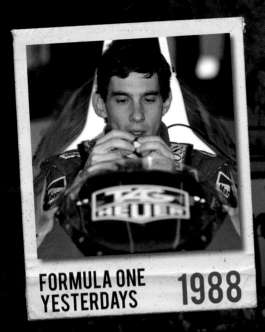

1988

A TWO-HORSE RACE FROM START TO FINISH

McLaren's engine contract with TAG (Porsche) had hardly wanted for success in the Formula One World Championship, bringing them three drivers' titles and two constructors' championships in four years, but when Honda became their supplier 25 years ago it sparked a season of unprecedented dominance.

Double world champion Alain Prost encouraged McLaren team boss Ron Dennis to sign the Brazilian Ayrton Senna, who had finished 3rd with Lotus in 1987, and between them the pair won 15 of the 16 races on the 1988 programme, occupying first and second places 10 times. McLaren accumulated 199 points to win the constructors' prize – a staggering 134 points more than their nearest pursuer and only two points fewer than the rest of the field combined.

The only race in which neither finished in the first two was the Italian Grand Prix at Monza, in which Prost retired with an engine misfire he had defied for two thirds of the race and Senna led until colliding with a tailender he was attempting to lap just two laps from the end. The incident left the Ferrari's of Gerhard Berger and Michele Alboreto to claim a 1-2 for the Italian team in the first Italian GP since the death of Enzo Ferrari.

Prost ended the season with more points (105) than Senna (94) but the title was decided on each driver's best 11 races, on which basis Senna's eight wins, including a superb drive in atrociously wet conditions in the British GP at Silverstone, trumped Prost's seven.

FACTS & STATS

- Michael Schumacher (Germany) holds the record for the most Formula One world drivers' championships with seven, followed by Juan Manuel Fangio (Argentina) with five and Alain Prost (France) with four.

- No country has produced more Formula One world champions than Great Britain, with 14, followed by Germany (9) and Brazil (8).

- The legendary Ferrari marque holds the record for the most Formula One constructors' titles with 15. Next come McLaren on 12 followed by Williams with seven.

- No driver has contested more Formula One Grand Prix races than the Brazilian Rubens Barrichello, with 322.

- The most successful driver in terms of Grand Prix victories is Michael Schumacher, who has stood on the winner's podium 91 times.

- Spain's Jaime Alguersuari became the youngest driver to start a Formula One Grand Prix when he raced in Hungary in 2009 at the age of 19 years 125 days.

- Sebastian Vettel of Germany is the youngest driver to win a Grand Prix, having been 21 years 73 days old when he won in Italy in 2008.

- There has been no older starter in a Grand Prix than Monaco-born Louis Chiron, who was 55 years 292 days when he drove a Maserati in the very first Formula One Grand Prix at Silverstone in 1950.

- Italy's Luigi Fagioli is credited with the record for the oldest winning driver in Formula One at 53 years 22 days, although he actually finished 11th in the race, the 1951 French GP, which was won by Juan Manuel Fangio, who switched to Fagioli's car after his own developed problems.

- Michael Schumacher holds the records for the most fastest laps (76) and the most pole positions (68) in Formula One history.

WHAT IS
GP2?

CONCEIVED BY BERNIE ECCLESTONE AND FLAVIO BRIATORE, GP2 WAS DESIGNED TO PROVIDE THE PERFECT PREPARATION FOR LIFE IN FORMULA ONE AND WAS INTRODUCED IN 2005, REPLACING FORMULA 3000 AS THE FEEDER SERIES FOR FORMULA ONE.

All teams in GP2 must use the same Dallara chassis, 4.0-litre Renault V8 engine and Pirelli tyres so that true driver ability is reflected. Two GP2 races – one of 180 kilometres and a 'sprint' event over 120km – are staged as support events at Formula One weekends so that drivers can also experience the Grand Prix environment.

Lewis Hamilton was GP2 champion in 2006 – two years before he won the Formula One title – and other past GP2 champions include Nico Rosberg (2005) and Romain Grosjean (2011). Six of the seven GP2 champions so far were competing in Formula One in 2012. When Charles Pic made his debut for Marussia in 2012 he became the 19th GP2 driver to step up to Formula One.

Here are four GP2 stars to watch out for.

JAMES CALADO
GREAT BRITAIN

Supported by the Racing Steps Foundation, 23-year-old Worcestershire-born James Calado has a strong CV that includes two Formula Renault 2.0 Winter Series titles in 2008, runner-up spot in the full Formula Renault 2.0 series in 2009 and second place in the prestigious British Formula Three championship of 2010, won by Frenchman Jean-Eric Vergne, who drove for Toro Rosso in Formula One in 2012.

Calado won five F3 races for Carlin before moving to Lotus in 2011, when he finished second in the GP3 series and made his GP2 debut. His maiden GP2 race victory came in the sprint event in last season's opening round in Malaysia.

ESTEBAN GUTIERREZ
MEXICO

Team-mate of James Calado at Lotus in 2012, Gutierrez competed in GP2 for the first time in 2011 alongside French driver Jules Bianchi and notched his first success when he won the sprint race at Valencia, having scored his first GP2 points by finishing 7th in the feature race the same weekend.

The youngest Mexican driver to win an international championship when he won the Formula BMW Europe series at 17 in 2008, 21-year-old Gutierrez scored back-to-back GP2 victories in the feature races at Valencia and Silverstone in 2012, while also testing in Formula One as reserve driver for Sauber.

JULES BIANCHI
FRANCE

Jules Bianchi competed for Tech 1 Racing in the Formula Renault 3.5 series in 2012 but finished 3rd in the 2010 and 2011 GP2 series for ART and was hired by Force India as reserve driver in last season's Formula One championship line-up, headed by British driver Paul di Resta and the 2009 GP2 champion Nico Hulkenberg.

From a distinguished French motor racing family – his grandfather Mauro was three times GT world champion while his uncle Lucien won the Le Mans 24 Hours and was 3rd in the 1968 Monaco Formula One Grand Prix – 23-year-old Jules was French Formula Renault 2.0 in 2007 in his debut season.

LUIZ RAZIA
BRAZIL

A seasoned GP2 campaigner, Razia scored his first victory in Bahrain at the end of the 2008-09 Asia GP2 series and won the sprint race at Monza in Italy in the European season that followed, but it was racing for Arden International in 2012 that he made his biggest impact to date, opening the season by winning the feature race in Malaysia and following up with sprint race success at Barcelona, Valencia and Silverstone to lead the series after 14 events.

Now 23, Razia was only 17 when he won the South American Formula Three title in 2006, with podium finishes in 11 of the 16 races and seven victories. He will hope his GP2 success opens more doors into Formula One, having tested for Virgin in 2010 and Team Lotus in 2011, when he was also reserve driver.

17 INDIA

VENUE: Buddh International Circuit, Uttar Pradesh

CIRCUIT LENGTHS: 5.125km

TURNS: 16

LAPS: 60

18 ABU DHABI

VENUE: Yas Marina Circuit

CIRCUIT LENGTHS: 5.554km

TURNS: 21

LAPS: 55

SITUATED IN THE NEW DELHI SUBURB OF NOIDA, THE BUDDH CIRCUIT MADE A SUCCESSFUL DEBUT IN 2011 AND WAS AN INSTANT HIT WITH FORMULA ONE DRIVERS.

Another new circuit designed by German architect Hermann Tilke, the track has two sizeable straights as opposed to the more conventional one and possesses an interesting mix of corners which throw up challenges for engineers and drivers.

The circuit's main feature is the multi-apex turn 10-11-12 sequence, which bears a resemblance to the famous turn 8 at Istanbul Park also designed by Tilke although the Buddh sequence arguably provides more overtaking chances.

Red Bull's Sebastian Vettel was at his irrepressible best when he won the inaugural Indian Grand Prix in 2011 by finishing comfortably clear of Britain's Jenson Button. In keeping with the local feel to the event, Indian cricket legend Sachin Tendulkar was charged with waving the chequered flag at the end of the race.

DID YOU KNOW?

In 2004, long before plans to build the Buddh circuit were formulated, locations in Mumbai and Hyderabad were strongly considered as possible future venues for a Formula One race in India.

A RELATIVELY RECENT ADDITION TO THE CALENDAR, DEBUTING IN 2009, THE ABU DHABI GRAND PRIX HAS BECOME ESTABLISHED AS AN ANNUAL DAY-NIGHT INSTALLMENT OF FORMULA ONE.

Certainly a region on an upward curve, Abu Dhabi has radically redeveloped on the back of their vast oil riches and one example of this is the new Yas Marina racing track.

In addition to the day-night novelty, the Abu Dhabi race also takes place amidst the stunning backdrop of Yas Island, with the track twisting its way round the island's glamorous marina.

The track is far from an improvised street circuit though. It was designed with racing in mind, with the longest straight on the Formula One calendar contributing to high speed sections while slow and precise turns provide frequent overtaking opportunities.

Sebastian Vettel won the first two races in Abu Dhabi, both of which were the last of the Formula One season. His second triumph in 2010 secured him that year's driver's World Championship, his first.

DID YOU KNOW?

A stone's throw from the Abu Dhabi track is Ferrari World, an amusement park owned by the Italian car giants that includes four theme parks, two water parks, 23 hotels and five golf courses.

19 UNITED STATES

VENUE: Circuit of the Americas, Austin, Texas

CIRCUIT LENGTHS: 5.516km

TURNS: 20

LAPS: 56

20 SPAIN

VENUE: Valencia Street Circuit

CIRCUIT LENGTHS: 5.419km

TURNS: 25

LAPS: 57

THE UNITED STATES GRAND PRIX RETURNED TO THE CALENDAR LAST YEAR, ITS REINSTATEMENT AFTER FIVE YEARS MARKED BY A CHANGE IN LOCATION, WITH THE NEW CIRCUIT OF THE AMERICAS IN TEXAS REPLACING THE INDIANAPOLIS SPEEDWAY.

The circuit, which hosted last year's penultimate race, took advantage of natural inclines to include elevation changes of up to 40m. Turn one, which consists of a steep uphill straight into a hairpin turn, is the stand-out feature.

Over 120,000 fans can attend races in Texas, using a combination of grandstands, temporary structures and natural seating areas, while the site of the track is intended also to house a medical facility, some meeting suites, conference facilities and banquet halls.

While the track is new, the US Grand Prix was first a part of Formula One in 1959 and has been held in California, Florida, New York and Arizona as well as Indianapolis.

The new track, which will host a MotoGP race in 2013, has hosted only one Formula One weekend before last year but Michael Schumacher has the most US Grand Prix victories at any venue with five. McLaren's Lewis Hamilton won the last race in Indianapolis in 2007.

DID YOU KNOW?

Hermann Tilke is the mastermind behind the track layout at the Circuit of the Americas. The German engineer has now designed nine new Formula One tracks and has been involved with redesigning six others.

HAVING HOSTED THE EUROPEAN GRAND PRIX FROM 2008 TO 2012, THE VALENCIA STREET CIRCUIT WILL RETAIN A REGULAR POSITION IN THE FORMULA ONE CALENDAR BY SHARING THE RESPONSIBILITY OF STAGING THE SPANISH GRAND PRIX WITH BARCELONA FROM 2013 ONWARDS.

Designed by prolific German architect Hermann Tilke, Valencia's street circuit first staged a Formula One race in 2008 which was won by Ferrari's Felipe Massa and is in a prime waterside location, with a section of the track even crossing over a canal.

The track combines fast straights with challenging S bends but was built with safety in mind; the track is at least 14 meters wide throughout and features adequate run-off areas on every corner.

Sebastian Vettel is the only multiple Valencia winner winning in both his Championship years of 2010 and 2011 while Venezuelan Williams driver Pastor Maldonado claimed his maiden Grand Prix victory.

DID YOU KNOW?

When Felipe Massa won the inaugural Valencia race, the Brazilian became the first man in the history of the sport to win his 100th race.

BORN: April 15, 1982; São Paolo, Brazil

2012 TEAM: Williams

DEBUT: 2010 Bahrain GP (HRT)

BEST FORMULA ONE RACE: 6th in 2012 Malaysia GP (Williams)

BEST FORMULA ONE SEASON: 18th in 2011 (Renault)

BRUNO SENNA

GIVEN THE TRAGEDIES IN HIS FAMILY HISTORY, IT IS PROBABLY NOT SURPRISING THAT BRUNO SENNA'S CAREER HAS NOT FOLLOWED A NORMAL PATTERN.

The nephew of the legendary Ayrton Senna, he began karting when he was five but when his father, Flavio, died in a motorcycle crash just two years after Ayrton's death at Imola in 1994 his mother insisted Bruno gave up his racing dream. It was almost 10 years before he resumed his career but his progress was rapid.

After just five races in the British Formula BMW series he stepped up to British Formula Three and was competing in GP2 just a year after that, finishing runner-up in the 2008 championship.

He tested in Formula One for Honda in the same year and looked set to be second driver to Jenson Button in 2009 only for Honda to withdraw. It meant Senna had to wait another year for his Formula One debut. That came with the Campos team, later to become HRT, in Bahrain in 2010.

After a season without a point in the slowest car on the grid he was not retained but, named as 2011 reserve driver by Renault, was back racing at the Belgian GP in August in place of Nick Heidfeld. He finished 9th in Italy but they were his only points and again he was dropped. Landing a 2012 drive at Williams, where he replaced Barrichello, Senna scored in four of his first eight races.

BORN: March 9, 1985; Maracay, Venezuela

2012 TEAM: Williams

DEBUT: 2011 Australian GP (Williams)

FIRST FORMULA ONE WIN: 2012 Spanish GP (Williams)

BEST FORMULA ONE SEASON: 19th in 2011 (Williams)

PASTOR MALDONADO

ONLY THE FOURTH VENEZUELAN IN FORMULA ONE – AND NOW THE FIRST TO WIN A GRAND PRIX – MALDONADO WON FOUR REGIONAL AND THREE NATIONAL KARTING CHAMPIONSHIPS BEFORE HE WAS 14.

His first experience of formula racing came in the Italian Formula Renault series, which he won at the second attempt in 2004 with eight race victories. He also finished 8th in the highly competitive European Formula Renault V.6 series the same year.

After testing for Minardi Formula One, he had a less successful 2005 but finished 3rd in the 2006 Formula Renault 3.5 series, which secured him a seat in GP2 for 2007 with Trident. He won impressively in Monaco in only his fifth race and finished the season in a respectable 11th place, despite missing four races with a fractured collarbone. Fifth with Piquet Racing in 2008, sixth in 2009 with ART, he hoped for a Formula One seat in 2010 but had to content himself with another year in GP2, albeit a very successful one.

Winning six races for Rapax, he took the title by a 16-point margin. It was enough for Williams to offer him their second race seat for 2011. He picked up only one point from his debut season but with a much improved Renault-powered car he scored his first victory from pole at the 2012 Spanish GP, giving Williams their first race win for eight years.

FORMULA ONE
YESTERDAYS   **1993**

1993

ALAIN PROST
4ème
TITRE

THE PROFESSOR HOLDS
HIS FINAL MASTERCLASS

Alain Prost, the French driver known as 'The Professor' for his intellectual air and meticulous planning, sits behind only Michael Schumacher and Juan Manuel Fangio in the pantheon of racing drivers. His four world titles have been bettered only by those two legends of the track. Until Schumacher passed it in 2001, his total of 51 was the most Grand Prix victories by any driver.

Champion in 1985, 1986 and 1989 for McLaren, Prost quit the team after the last of those three wins as his intense rivalry with team-mate Ayrton Senna descended into bitterness. It continued during Prost's first season with Ferrari, when for the second year running a collision between the two in the penultimate race in Japan effectively determined the title.

Prost took a year off in 1992 but returned with Williams in 1993 and comprehensively beat the Brazilian, who won five times in an inferior car but could not overhaul the lead Prost built by winning seven of the first 10 races.

He clinched the title in the 14th of the 16 races, in Portugal, by finishing second to Michael Schumacher, with Senna failing to finish after his engine blew. He had already announced his retirement at the end of the season following the news that Senna would be driving for Williams in 1994.

There appeared to be no love lost between the rivals, yet when Senna beat Prost in the final race of the season, the Brazilian embraced the Frenchman on the podium, taking Prost aback. After Senna was killed the following year, Prost was honoured to be a pallbearer at his funeral.

SPOTLIGHT ON SILVERSTONE

Silverstone is the home of British motorsport and its association with the FIA Formula One World Championship dates back to the competition's beginning in 1950.

The current venue for the British Grand Prix hosted the very first world championship race on May 13, 1950 won by eventual champion Giuseppe Farina.

The circuit, on the Northamptonshire and Buckinghamshire border, has been the permanent home for the race since 1987, having shared the responsibility with first Aintree and then Brands Hatch from 1955.

British drivers have had mixed success at Silverstone, with the first world championship win at the venue for a Brit provided by Peter Collins in 1958.

The 1960s were a golden period for British racing and Jim Clark led a British 1-2-3 in 1963, finishing ahead of John Surtees and Graham Hill. Clark won again at Silverstone two years on, this time incredibly with British drivers filling the first five places.

Silverstone was renowned for being one of the fastest tracks in the calendar until it was modified in 1991, and accidents were common. In one notable crash in

1973, Jody Scheckter who won the race a year later spun at the end of the first lap, taking eight cars out in the process.

In 1997, Jacques Villeneuve surged to his second straight Silverstone after two race leaders had been forced to retire; Michael Schumacher led by 40 seconds when a wheel bearing fault ended his race and then Mika Häkkinen's engine failed as he led the field with only seven laps left.

Schumacher, a three-time winner, will also be remembered for sealing his 1998 victory in the pit lane while serving a penalty, as well as forcing Fernando Alonso off the track in 2003. The race that year is also remembered for the controversial Irish protester Neil Horan – a former priest – invading the track on the Hangar straight.

Silverstone is often unlucky with the weather, with rain regularly making life miserable for spectators, but often it adds an extra challenge to the racing. Ayrton Senna's sole Silverstone victory in 1988 came in pouring rain while Lewis Hamilton, another driver with particular skill in inclement conditions, won on a wet track there en route to his 2008 title.

Year	Driver	Constructor	Year	Driver	Constructor
1950	Giuseppe Farina (Italy)	Alfa Romeo	1981	Nelson Piquet (Brazil)	Brabham
1951	Juan Manuel Fangio (Argentina)	Alfa Romeo	1982	Keke Rosberg (Finland)	Williams
1952	Alberto Ascari (Italy)	Ferrari	1983	Nelson Piquet (Brazil)	Brabham
1953	Alberto Ascari (Italy)	Ferrari	1984	Niki Lauda (Austria)	McLaren
1954	Juan Manuel Fangio (Argentina)	Maserati Mercedes	1985	Alain Prost (France)	McLaren
1955	Juan Manuel Fangio (Argentina)	Mercedes	1986	Alain Prost (France)	McLaren
1956	Juan Manuel Fangio (Argentina)	Ferrari	1987	Nelson Piquet (Brazil)	Williams
1957	Juan Manuel Fangio (Argentina)	Maserati	1988	Ayrton Senna (Brazil)	McLaren
1958	Mike Hawthorn (Great Britain)	Ferrari	1989	Alain Prost (France)	McLaren
1959	Jack Brabham (Australia)	Cooper	1990	Ayrton Senna (Brazil)	McLaren
1960	Jack Brabham (Australia)	Cooper	1991	Ayrton Senna (Brazil)	McLaren
1961	Phil Hill (United States)	Ferrari	1992	Nigel Mansell (Great Britain)	Williams
1962	Graham Hill (Great Britain)	BRM	1993	Alain Prost (France)	Williams
1963	Jim Clark (Great Britain)	Lotus	1994	Michael Schumacher (Germany)	Benetton
1964	John Surtees (Great Britain)	Ferrari	1995	Michael Schumacher (Germany)	Benetton
1965	Jim Clark (Great Britain)	Lotus	1996	Damon Hill (Great Britain)	Williams
1966	Jack Brabham (Australia)	Brabham	1997	Jacques Villeneuve (Canada)	Williams
1967	Denny Hulme (New Zealand)	Brabham	1998	Mika Häkkinen (Finland)	McLaren
1968	Graham Hill (Great Britain)	Lotus	1999	Mika Häkkinen (Finland)	McLaren
1969	Jackie Stewart (Great Britain)	Matra	2000	Michael Schumacher (Germany)	Ferrari
1970	Jochen Rindt (Austria)	Lotus	2001	Michael Schumacher (Germany)	Ferrari
1971	Jackie Stewart (Great Britain)	Tyrell	2002	Michael Schumacher (Germany)	Ferrari
1972	Emerson Fittipaldi (Brazil)	Lotus	2003	Michael Schumacher (Germany)	Ferrari
1973	Jackie Stewart (Great Britain)	Tyrell	2004	Michael Schumacher (Germany)	Ferrari
1974	Emerson Fittipaldi (Brazil)	McLaren	2005	Fernando Alonso (Spain)	Renault
1975	Niki Lauda (Austria)	Ferrari	2006	Fernando Alonso (Spain)	Renault
1976	James Hunt (Great Britain)	McLaren	2007	Kimi Räikkönen (Finland)	Ferrari
1977	Niki Lauda (Austria)	Ferrari	2008	Lewis Hamilton (Great Britain)	McLaren
1978	Mario Andretti (United States)	Lotus	2009	Jenson Button (Great Britain)	Brawn
1979	Jody Scheckter (South Africa)	Ferrari	2010	Sebastian Vettel (Germany)	Red Bull
1980	Alan Jones (Australia)	Williams	2011	Sebastian Vettel (Germany)	Red Bull

ROLL OF HONOUR

MULTIPLE WINNERS	Michael Schumacher (Germany)	7
	Juan Manuel Fangio (Argentina)	5
	Alain Prost (France)	4
WINNERS BY COUNTRY	Great Britain	14
	Germany	9
	Brazil	8
WINNERS BY CONSTRUCTOR	Ferrari	15
	McLaren	12
	Williams	7
MOST GPS CONTESTED	Rubens Barrichello (Brazil	322
MOST GPS WON	Michael Schumacher (Germany)	91
YOUNGEST STARTER	Jaime Algersuari (Spain)	19 years 125 days
YOUNGEST WINNER	Sebastian Vettel (Germany)	21 years 73 days
OLDEST STARTER	Louis Chiron (Monaco)	55 years 292 days
OLDEST WINNER	Luigi Fagioli (Italy)	53 years 22 days
MOST FASTEST LAPS IN CAREER	Michael Schumacher (Germany)	76
MOST POLE POSITIONS	Michael Schumacher (Germany)	68

Year	Constructor	Country	Drivers
1958	Vanwall	GB	Stirling Moss, Tony Brooks
1959	Cooper	GB	Jack Brabham, Moss, Bruce McLaren
1960	Cooper	GB	Brabham, McLaren
1961	Ferrari	Italy	Phil Hill, Wolfgang von Trips
1962	BRM	GB	Graham Hill
1963	Lotus	GB	Jim Clark
1964	Ferrari	Italy	John Surtees, Lorenzo Bandini
1965	Lotus	GB	Clark
1966	Brabham	GB	Brabham
1967	Brabham	GB	Denny Hulme, Brabham
1968	Lotus	GB	Hill, Jo Siffert, Clark, Jackie Oliver
1969	Matra	France	Jackie Stewart, Jean-Pierre Beltoise
1970	Lotus	GB	Jochen Rindt, Emerson Fittipaldi, Hill,John Miles
1971	Tyrrell	GB	Stewart, François Cevert
1972	Lotus	GB	Fittipaldi
1973	Lotus	GB	Fittipaldi, Ronnie Peterson
1974	McLaren	GB	Fittipaldi, Hulme, Mike Hailwood, David Hobbs, Jochen Mass
1975	Ferrari	Italy	Clay Regazzoni, Niki Lauda
1976	Ferrari	Italy	Lauda, Regazzoni
1977	Ferrari	Italy	Lauda, Carlos Reutemann
1978	Lotus	GB	Mario Andretti, Peterson
1979	Ferrari	Italy	Jody Scheckter, Gilles Villeneuve
1980	Williams	GB	Alan Jones, Reutemann
1981	Williams	GB	Jones, Reutemann
1982	Ferrari	Italy	Villeneuve, Didier Pironi, Patrick Tambay, Andretti
1983	Ferrari	Italy	Tambay, René Arnoux
1984	McLaren	GB	Alain Prost, Lauda
1985	McLaren	GB	Lauda, Prost, Watson
1986	Williams	GB	Nigel Mansell, Nelson Piquet
1987	Williams	GB	Mansell, Piquet
1988	McLaren	GB	Prost, Ayrton Senna
1989	McLaren	GB	Senna, Prost
1990	McLaren	GB	Senna, Gerhard Berger
1991	McLaren	GB	Senna, Berger
1992	Williams	GB	Mansell, Riccardo Patrese
1993	Williams	GB	Damon Hill, Prost
1994	Williams	GB	Hill, Senna, David Coulthard, Mansell
1995	Benetton	GB	Michael Schumacher, Johnny Herbert
1996	Williams	GB	Hill, Jacques Villeneuve
1997	Williams	GB	Villeneuve, Heinz-Harald Frentzen
1998	McLaren	GB	Coulthard, Mika Häkkinen
1999	Ferrari	Italy	Schumacher, Eddie Irvine, Mika Salo
2000	Ferrari	Italy	Schumacher, Rubens Barrichello
2001	Ferrari	Italy	Schumacher, Barrichello
2002	Ferrari	Italy	Schumacher, Barrichello
2003	Ferrari	Italy	Schumacher, Barrichello
2004	Ferrari	Italy	Schumacher, Barrichello
2005	Renault	France	Fernando Alonso, Giancarlo Fisichella
2006	Renault	France	Alonso, Fisichella
2007	Ferrari	Italy	Felipe Massa, Kimi Räikkönen
2008	Ferrari	Italy	Räikkönen, Massa
2009	Brawn	GB	Jenson Button, Barrichello
2010	Red Bull () ()	Austria	Sebastian Vettel, Mark Webber
2011	Red Bull ()()	Austria	Vettel, Webber

BORN: October 19, 1981; Suomussalmi, Finland

2012 TEAM: Caterham

DEBUT: 2007 Australian GP (Renault)

BEST FORMULA ONE WIN: 2008 Hungarian GP (McLaren)

BEST FORMULA ONE SEASON: 7th in 2007 (Renault) and 2008 (McLaren)

HEIKKI KOVALAINEN

KOVALAINEN SET HIS SIGHTS ON A CAREER BEHIND THE WHEEL AS YOUNG AS 10 YEARS OLD AND AFTER MAKING A NAME FOR HIMSELF IN KARTING HE RACED CARS FOR THE FIRST TIME IN THE 2001 BRITISH FORMULA RENAULT SERIES, IN WHICH HE WAS 'ROOKIE OF THE YEAR' WITH TWO DEBUT-SEASON WINS.

Promptly signed to Renault's driver development programme, he moved quickly into British Formula Three, again making an immediate impression, winning five times in his first season to finish 3rd. Another step up brought his first career title in the 2004 World Series by Nissan.

The success led to him testing for Renault Formula One for the first time and when he defeated Michael Schumacher in the 2004 Race of Champions it seemed he was destined for a bright future. In the event, he had to wait until 2007 for his first Formula One race seat even after finishing runner-up in the inaugural GP2 series in 2005.

A solid debut for the defending champions was followed by an ever better second season, this time alongside Lewis Hamilton at McLaren. Kovalainen landed his first GP win in Hungary and finished 7th in the standings. Unfortunately, he was dropped at the end of 2009 in favour of new world champion Jenson Button and has since been with Team Lotus – since rebranded as Caterham – he has been praised for his work helping the new team make year-on-year improvement.

BORN: September 8, 1984; Vyborg, Russia

2012 TEAM: Caterham

DEBUT: 2010 Bahrain GP (Renault)

BEST FORMULA ONE RACE: 23rd in 2011 Australian GP (Lotus Renault)

BEST FORMULA ONE SEASON: 10th in 2011 (Lotus Renault)

VITALY PETROV

VITALY PETROV REACHED FORMULA ONE VIA A LONG ROUTE, BEGINNING WITH ICE RACING EVENTS AND RALLY SPRINTS, IN WHICH HE BECAME RUSSIAN CHAMPION.

In 2002 he won every race in the Russian Lada Cup series, before moving to Europe to contest the Italian and European Formula Renault series. Back home, he won the Russian Sports Car and Russian Formula 1600 titles in 2005, then returned to Europe to finish 3rd in both the Euroseries 3000 and Italy's F3000 series. He then spent three seasons in GP2, claiming his first win with Campos at Valencia in 2007, finishing 13th overall, and improving to 7th the following year, when he won in Valencia again, on the street circuit.

In 2009, switching to Barwa Addax, he won for a third time in Valencia as well as in Turkey, finishing runner-up to Nico Hulkenberg. In 2010, finally, Petrov was given a chance in Formula One with Renault as the first Russian to compete at that level. His debut season brought some excellent results, notably 5th in Hungary and 6th in Abu Dhabi, to finish 13th overall.

His second season began with a first podium finish in Australia but after finishing 10th in the standings Renault dropped him for 2012. It was not until a month before the new season that Caterham gave him the chance to continue his Formula One career.

BORN: April 16, 1986; Uphall, Scotland

2012 TEAM: Force India

DEBUT: 2011 Australian GP (Force India)

BEST FORMULA ONE RACE: 6th in 2011 Singapore GP and 2012 Bahrain GP (both Force India)

BEST FORMULA ONE SEASON: 13th in 2011 (Force India)

PAUL DI RESTA

EVIDENCE OF THE PROGRESS MADE BY PAUL DI RESTA AFTER ONLY TWO YEARS IN FORMULA ONE WAS THAT IT TOOK HIM ONLY THE FIRST EIGHT RACES OF THE 2012 SEASON TO MATCH HIS POINTS TOTAL FOR THE WHOLE OF 2011.

Born in Scotland of Italian heritage – the Scottish-born Indycar champion Dario Franchitti is his cousin – Di Resta switched from kart to car racing in 2002, finishing 3rd in the UK's Formula Renault series in 2004, which saw him named McLaren Autosport BRDC Young Driver of the Year.

Next came the Formula Three Euroseries, where he finished 10th in 2005 before taking the title in 2006 with ASM, winning five times against a field that included future Formula One world champion Sebastian Vettel. With financial support from Mercedes-Benz, Di Resta spent the next four seasons in the ultra-competitive DTM touring car championship in Germany, finishing 5th, 2nd and 3rd before clinching the title with HWA in 2010 after three race wins. In the meantime he was making good progress testing in Formula One with Force India, making a big impression in eight Friday test sessions in 2010, achieving excellent times.

It was confirmed he would race alongside Sutil in 2011 and after a sixth place in Singapore and 13th place in the driver table he was the clear choice for 'rookie of the year' and was retained by Force India for 2012.

BORN: August 19, 1987; Emmerich am Rhein, Germany

2012 TEAM: Force India

DEBUT: 2010 Australian GP (Williams)

BEST FORMULA ONE RACE: 5th in 2012 European GP (Force India)

BEST FORMULA ONE SEASON: 14th in 2011 (Williams)

NICO HULKENBERG

WHEN NICO HULKENBERG WON THE 2003 GERMAN KARTING CHAMPIONSHIP HE WAS EMULATING MICHAEL SCHUMACHER 16 YEARS EARLIER.

With Schumacher's manager, Willi Weber, looking after his affairs, he made a spectacular debut aged just 18 in German Formula BMW in 2005, comfortably winning the title with eight wins, following in the footsteps of another German high-flyer, Sebastian Vettel, who had won the series only the year before.

In 2006 he led Team Germany to the A1GP title, winning nine races. He also raced in German Formula Three that year, winning once, and in 2007 moved up another level in the F3 Euroseries, winning four races to finish 3rd overall.

An impressive Formula Masters victory was followed by an invitation to test for Williams Formula One, for whom he secured a deal as test driver in 2008, while at the same time taking the F3 Euroseries title with seven wins.

The next year he raced in GP2, where again he excelled, winning five times to be champion with a race to spare. His Formula One breakthrough came with Williams in 2010 but even though he was 10th in Malaysia and 6th in Hungary, he was dropped for 2011 in favour of Pastor Maldonado. Switching to Force India, he had to be content with a test and reserve role for his first season but landed a race seat for 2012 and confirmed his promise with a first top-six placing in the European GP in Valencia.

1950s

The cars that raced in Formula One in the 1950s were little changed from the pre-war years. Built using steel panels on a tubular frame, they were heavy cars with front-mounted engines but very little protection for the driver at speeds of up to 125mph.

Top left: British Grand Prix 1953 Silverstone July 1953 Mike Hawthorn in his Ferrari number 8 car at Copse corner.

Left: British Grand Prix 1953 Silverstone July 1953 The winner Alberto Ascari in his Ferrari number 5 car drives past the grandstand.

Bottom: The start of the 1954 British Grand Prix at Silverstone.

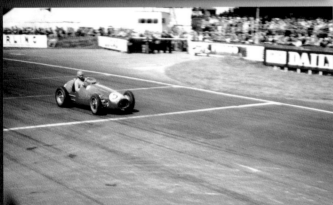

THE CAR'S THE STAR

THE CHANGING SHAPE OF FORMULA ONE CARS

1960s

The second decade of Formula One featured cars with a sleeker design. The aluminium monocoque chassis developed for the iconic Lotus 25 revolutionised cars by making them lighter and faster but more rigid, and giving more protection to the driver.

British Grand Prix 1965 Silverstone July 1965 John Surtees sits in his Ferrari number 1 car.

Motor racing 1962 at Aintree Liverpool Action from the British Grand Prix. Jim Clark in the Lotus 20 leads Carel Godin de Beaufort.

1970s

With the addition of wings and aerofoils and wider tyres, the racing cars of the 1970s were more efficient and held the road better and in appearance began to resemble today's cars.

Formula One 1975 Monaco Grand Prix, 11/5/75 Emerson Fittipaldi, McLaren in race action.

Formula One 1970 Mexican Grand Prix, 25/10/70 Jacky Ickx, Ferrari.

Formula One San Marino Grand Prix Imola 5/5/85 Keke Rosberg, Williams.

1980s

The era of the turbo-charged car peaked in 1985 when Keke Rosberg in a Williams-Honda became the first driver to lap the Silverstone circuit at 160mph during a practice session.

Ayrton Senna, Brazil, Mclaren.

1990s

After the turbo era, designers looked for other ways to gain an edge and it was the lowline chassis pioneered by McLaren – more aerodynamically efficient and generating more downforce – that enabled Ayrton Senna to be world champion three times.

Japanese Grand Prix, Suzuka, 20/10/91 Ayrton Senna, McLaren.

2000s

The Formula One cars of the new millennium race at speeds up to 200mph and can go faster still. The Ferrari F2002 that Michael Schumacher drove to the 2002 world title is reckoned by some to be the fastest Formula One car of all time. The F2012 driven by Fernando Alonso in last year's championship is notable for its stepped nose.

Ferrari Formula One driver Fernando Alonso of Spain drives during the German Formula One Grand Prix at the Hockenheimring in Hockenheim July 22, 2012.

Ferrari driver Michael Schumacher crosses the line to win the race Japan Grand Prix Suzuka.

Michael Schumacher Ferrari Brazil Grand Prix Interlagos, Sao Paulo Sunday 31st March 2002.

BORN: March 18, 1982; Lindenfels, Germany

2012 TEAM: Marussia

DEBUT: 2004 Canadian GP (Jordan)

BEST FORMULA ONE RACE: 2nd in 2008 Hungarian GP and 2009 Singapore GP (both Toyota)

BEST FORMULA ONE SEASON: 10th in 2008 and 2009 (both Toyota)

TIMO GLOCK

TIMO GLOCK MADE HIS FORMULA ONE DEBUT IN ONLY HIS FIFTH COMPETITIVE SEASON BUT HAD TO WAIT ALMOST FOUR MORE YEARS FOR HIS FORMULA ONE CAREER TO BEGIN IN EARNEST.

A late starter in karting at 15, he was racing cars after only three years. Junior Cup winner on his debut season in Formula BMW-ADAC, he won the main title in 2001 and stepped up to German Formula Three in 2002, winning three races for Opel to finish 3rd in the standings before finishing 5th in the Formula Three Euroseries in 2003 after three more wins.

Snapped up by Jordan Formula One as their third driver for 2004, he came home a stunning 7th on his debut in Canada and had a further three races that season. With no permanent seat, however, he spent 2005 in the Champ Car series in America and on returning to Europe spent two years in GP2.

The second of those brought him the GP2 title, which was enough to tempt Toyota to sign him as Ralf Schumacher's replacement for 2008. Despite a number of accidents, he took 10th place in the standings in both 2008 and 2009. Unfortunately for him, Toyota withdrew from Formula One and while a move to Virgin – now Marussia – kept him racing it was in a much less competitive car and he failed to pick up a point in his first two seasons.

CHARLES PIC

BORN: February 15, 1990; Montelimar, France

2012 TEAM: Marussia

DEBUT: 2012 Australian GP (Marussia)

BEST FORMULA ONE RACE: 15th in 2012 Australian and European GPs (both Marussia)

BEST FORMULA ONE SEASON: 2012 (Marussia)

NO SURPRISE THAT FRENCHMAN CHARLES PIC HAS PURSUED A CAREER IN MOTOR RACING, HAVING BEEN ENCOURAGED TO TAKE TO THE TRACK BEFORE HE WAS 12 BY HIS GODFATHER, THE FORMER FORMULA ONE DRIVER ERIC BARNARD, WHO GAVE HIM A KART AS A PRESENT.

After some success in the junior categories, in 2006, he made his debut in the French Formula Campus championship, creating an immediate impression by finishing 3rd after two wins and was just as impressive in the Formula Renault 2.0 Eurocup in 2007, again finishing 3rd and scoring his first international race win in Germany.

In two seasons in the World Series by Renault Formula 3.5 championship he finished 6th and 3rd, finding time in 2009 to make his debut in GP2 and record his first win, for Arden in the Asia Series. He stepped up to the main GP2 series in 2010, again with Arden, claiming one victory, and the following year finished 4th overall after wins in Spain and Monaco.

Invited by Virgin to try out in Formula One, he performed well at the young driver test in Abu Dhabi in November 2011 and was signed to replace Jerome D'Ambrosio alongside Timo Glock in their 2012 line-up and made a creditable start that promised greater things once he has the chance to drive a more competitive car.

DRIVER PROFILES 22

PEDRO DE LA ROSA

BORN: February 24, 1971; 15, 1990; Cardedeu, Spain

2012 TEAM: HRT

DEBUT: 1999 Australian GP (Arrows)

BEST FORMULA ONE RACE: 2nd in 2006 Hungarian GP (McLaren)

BEST FORMULA ONE SEASON: 11th in 2006 (McLaren)

VETERAN PEDRO DE LA ROSA SCORED A POINT ON HIS FORMULA ONE DEBUT WHEN HE FINISHED SIXTH IN AUSTRALIA IN 1999 BUT THOUGH HE HAD CLOCKED UP 87 RACES AT THE BEGINNING OF THE 2012 SEASON HE WAS STILL LOOKING FOR HIS FIRST WIN.

A late starter in karting at 17, he enjoyed much success when he moved into cars, winning eight titles in his first nine seasons, including Spanish Formula Ford, Formula Renault Great Britain, All-Japan Formula Three, and All-Japan GT championships.

In 1998 he began testing in Formula One for Jordan and made his race debut the following season. He scored points again for Arrows in 2000 and for Jaguar in 2001 but missed out in 2002 and lost his seat. He spent the next seven years combining test duties with some racing for McLaren, enjoying his best season in 2006, when he collected 19 points from eight starts, including a 2nd in Hungary behind Jenson Button.

His reputation as a tester was by now considerable but he craved racing and returned to competition with BMW Sauber in 2010. It did not work out well and he was back in his testing role with McLaren in 2011, when his one race was as stand-in for Sauber's Sergio Perez in Canada, where he finished a creditable 12th, so it came as a surprise to many when it was announced he had landed a two-year race contract with Spanish team HRT.

NARAIN KARTHIKEYAN

BORN: January 14, 1977; Chennai, India

2012 TEAM: HRT

DEBUT: 2005 Australian GP (Jordan)

BEST FORMULA ONE RACE: 4th in 2005 United States GP (Jordan)

BEST FORMULA ONE SEASON: 18th in 2005 (Jordan)

INDIA'S FIRST FORMULA ONE DRIVER, KARTHIKEYAN IS THE SON OF A FORMER INDIAN RALLY CHAMPION WHO SENT HIM TO LEARN HIS CRAFT IN EUROPE.

A fast learner, in 1994 he became the first Indian to win the British Formula Ford Winter series and then the first from his country – indeed the first Asian – to win the Formula Asia series in 1996. Stepping up to British Formula Three he enjoyed year-on-year improvement, finishing 4th in 2000, and in 2001 he became the first Indian to test a Formula One car, with Jaguar at Silverstone.

However, he did not secure a Formula One race drive until 2005 after three seasons in the Formula Nissan World Series, finishing 4th in 2003. He tested in Formula One for Minardi in 2004 but it was Jordan who gave him his Formula One break.

Sadly, he had a disappointing debut season and while he finished fourth in the United States GP it was as one of only six starters after an argument over tyre safety. He lost the seat for 2006 and while he had a test role for Williams his racing career went down a different route, including three seasons in A1 GP, the Le Mans series in 2009 – when he became the first Indian to compete in the legendary 24 Hours race – and NASCAR Camping World Truck Series. But he was given another crack at Formula One with HRT in 2011, when his nine starts included the inaugural Indian GP.

DRIVER PROFILES 24

QUIZ ANSWERS

P25: WORDSEARCH

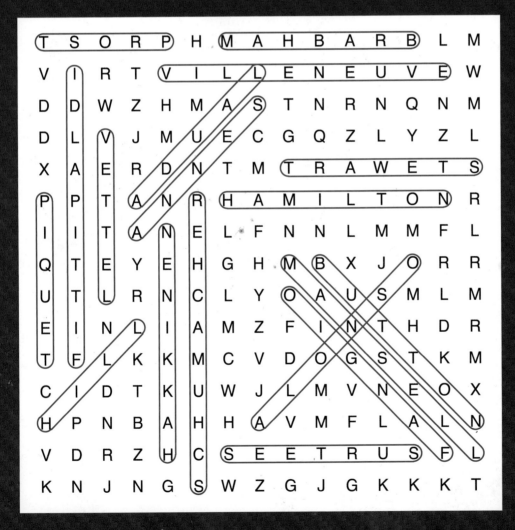